Grimm/Krummen

Stability Testing in the EC, Japan and the USA

Scientific and Regulatory Requirements
(including the draft of the Harmonised Guideline for the EC, Japan and the USA)

Paperback APV

Die Buchreihe wird im Auftrag der Arbeitsgemeinschaft für Pharmazeutische Verfahrenstechnik e.V. (APV), Mainz, herausgegeben.

In der Reihe Paperback APV sind bereits erschienen:

Band 1 Hauschild

Ausgangsmaterialien für die Arzneimittelherstellung. Qualitätsanforderungen (GMP) und deren Folgen für Zulieferer und Pharma-Betriebe.

Band 2 Essig/Schmidt/Stumpf

Flüssige Arzneimittelformen und Arzneimittelsicherheit. Unter besonderer Berücksichtigung der Qualität von Wasser

Band 3 Hasler

Dosierungsgenauigkeit einzeldosierter fester Arzneiformen

Band 4 Brandau/Lippold

Dermal and Transdermal Absorption

Band 5 Schrank/Skinner

Arzneimittelhygiene von der Herstellung bis zur Verarbeitung

Band 6 Oeser

Selbstinspektion und Inspektion. GMP-Forderungen zur Qualitätssicherung im Arzneimittelbereich

Band 7 Fahrig/Hofer

Die Kapsel. Grundlagen, Technologie und Biopharmazie einer modernen Arzneiform

Band 8 Sucker

Praxis der Validierung unter besonderer Berücksichtigung der FIP-Richtlinien für die gute Validierungspraxis

Band 9 Algner/Helbig/Spingler

Primär-Packmittel. Herstellung + Optimierung + Kontrolle = Sicherheit

Band 10 Asche/Essig/Schmidt

Technologien von Salben, Suspensionen und Emulsionen

Band 11 Hanke

Qualität pflanzlicher Arzneimittel. Prüfung und Herstellung – Anforderungen bei der Zulassung

Band 12 Dertinger/Gänshirt/Steinigen

GAP – Praxisgerechtes Arbeiten in pharmazeutisch-analytischen Laboratorien

Band 13 Helbig/Spingler

Kunststoffe für die Pharmazeutische Verpackung. Einsatz – Anforderungen – Prüfverfahren

Band 14 Brandau

Pharmaproduktion aus betriebswirtschaftlicher Sicht

Band 15 Essig/Hofer/Schmidt/Stumpf

Stabilisierungstechnologie.
Wege zur haltbaren Arzneiform

Band 16 Grimm

Stability Testing of Pharmaceutical Products

Band 17 Müller

Controlled Drug Delivery

Band 18 Schultz/Grünewald

Strahlen- und Gassterilisation

Band 19 Hauschild/Spingler

Migration bei Kunststoffverpackungen

Band 20 Altenschmidt/Hasler/Möller

Probenahme bei der Qualitätssicherung von Arzneimitteln

Band 21 Scheer

Der Limulustest

Band 22 Gundert-Remy/Möller

Oral Controlled Release Products

Band 23 Essig/Stumpf

Flüssige Arzneiformen schwerlöslicher Arzneistoffe

Paperback APV

Die Buchreihe wird im Auftrag der Arbeitsgemeinschaft für Pharmazeutische Verfahrenstechnik e.V. (APV), Mainz, herausgegeben.

Band 24 Elste

Haltbarkeit von Grundstoffen und Zubereitungen in der Apotheke

Band 25 Gurny/Junginger

Bioadhesion – Possibilities and Future Trends

Band 26 Kammerl/Verheyen

Just-in-Time. Theorie und Praxis in der Arzneimittelherstellung

Band 27 Moest/Werani

In-Prozeß-Kontrolle fester Arzneiformen

Band 28 Lingnau

Hygieneunterweisung für Mitarbeiter im Pharma-Betrieb

Band 29 Blume/Gundert-Remy/Möller

Controlled Modified Release Products

Band 30 Möller/Oeser

Drug Master Files. Global Harmonization of Quality Standards

Band 31 Gurny/Teubner

Dermal and Transdermal Delivery – New Inrights and Perspectives

Stability Testing in the EC, Japan and the USA

Scientific and Regulatory Requirements

(including the draft of the Harmonised Guideline for the EC, Japan and the USA)

Edited by

Dr. Wolfgang Grimm
Dr. Karl Thoma GmbH, D-Biberach
and
Dr. Kurt Krummen
Sandoz AG, CH-Basel

with 84 figures and 92 tables

WVG

Wissenschaftliche Verlagsgesellschaft mbH Stuttgart 1993

Ein Markenzeichen kann warenzeichenrechtlich geschützt sein, auch wenn ein Hinweis auf etwa bestehende Schutzrechte fehlt.

Die Deutsche Bibliothek – CIP-Einheitsaufnahme

Stability testing in the EC, Japan and the USA : scientific and regulatory requirements (including the draft of the harmonised guidelines for the EC, Japan and the USA) ; with 92 tables / ed. by Wolfgang Grimm and Kurt Krummen. – Stuttgart : Wiss. Verl.-Ges., 1993
(Paperback APV ; 32)
ISBN 3-8047-1215-0
NE: Grimm, Wolfgang [Hrsg.]; International Association for Pharmaceutical Technology: Paperback APV

Printed in Germany

Satz und Druck: Gulde-Druck GmbH, Tübingen

A Introduction

Stability Testing in the EC, Japan and the USA, Scientific and Regulatory Requirements

Wolfgang Grimm, D-Biberach/Riß

Stability testing is performed to ensure that drug products retain their full efficacy up to the end of their expiration date.

The results obtained are used to formulate the stability information, which ensures the quality, efficacy and safety of the drug product. This aim can be reached only by the concerted action of many branches of knowledge. Stability testing is a science based on pharmaceutical technology, analytics, physico-chemistry, organic-chemistry, plastic-chemistry, biochemistry, biotechnology and statistics.

Stability testing accompanies the development of a drug product from the experiments with the drug substance over formulation development up to the final dose and formula for introduction, but it continues during production and necessary major changes.

The whole field of stability testing must be planed very carefully. Most important is a systematic development in a logically structured and optimally coordinated system. By this means the state of knowledge can be continuously extended, the stability information is than based on all the results gained during development.

Pharmaceutical industry is performing stability testing due to the responsibility for the drug products.

The results are then checked carefully by the authorities for marketing authorisation. To standardize stability testing, rules have been worked out and published as guidelines.

Nearly all countries have their own guidelines. In the basic principles the guidelines show no decisive difference, but in the concept, more general, more detailed, and in formal requirements.

It must be the aim to elaborate the basic principles of stability testing and to demonstrate the formal differences. On the basis of this a harmonized guideline may be obtained accepted by as many countries as possible.

Then it will be not necessary any more, to perform parallel stability programs with only formal differences for an international registration. This would mean enormous savings in cost and labor for industry and the regulatory bodies.

After 1985 the APV has organized for a second time an International Symposium on Stability Testing which took place in Darmstadt June 5–7, 1991.

It was the aim

- to present the state of the art in stability testing and the newest official requirements in the EC, Japan and the USA
- to indicate future perspectives in stability testing
- to reach a harmonized guideline on stability testing for the EC, Japan and the USA and thereby covering 83% of the world market for drug products.

The paperback presents the results of the symposium, the lectures and the revised draft for a harmonized guideline on stability testing for the EC, Japan and the USA. Up to now, the guideline after acceptance will also be taken over by Canada and Australia, further countries may follow.

The paperback is subdivided into 4 chapters:

- Stability testing during development, state of the art.
- Official requirement of the EC, Japan and the USA and its application.
- Harmonized guideline for the stability testing of new active drug substances and products and its application.
- Future perspectives in stability testing.

The first chapter consists of 6 papers.

First the basic principles of stability testing are described, then all the necessary experiments for a drug substance are presented, the physical and chemical stability, the impact of polymorphism and the conclusions, the selection of the chemical form of the drug substance, physical and chemical stabilization.

The paper "Excipient compatability" gives a survey of known reactivities between a drug substance and excipients. It is recommended to study compacts of excipient blends with the drug at the desired potency level instead of binary mixtures. A promising technique to study solid state reactions is heat conducting calorimetry besides microcalorimetry.

Important for the evaluation of stress data and long term data are the new techniques and strategies for the prediction of the stability.

Non linear regression analysis and error propagation law are employed, the differences between the classical and the new methods are described and the conclusions drawn.

The most frequent analytical method applied in stability testing is the HPLC. In a critical review HPLC, SFC and CZE are compared.

Selection of the packaging material is an import part during development. This is covered by the paper "Influence of packaging material on the stability of drug products".

The second chapter which describes the official requirements begins with a paper "Plastics for drug products". Many experiments are described to find out interactions, migrations, sorptions, but also official requirements are mentioned, mainly for the EC.

Then the EC Guideline for stability tests on active substances and finished products follow. Besides the requirements also the philosophy behind is described.

The revised Japanese Guidelines for stability studies describe the newest requirements of Japan. The FDA Guideline is under revision. The considerations on the revision terminate the official requirements.

It follows a paper, stability testing in relations to Guidelines in the EC, Japan and the USA, a recommendation.

The third chapter begins with the newest draft of the harmonized guideline on stability testing. It is based on the Darmstadt Declaration elaborated by a concensus group headed by A.C. Cartwright (CPMP) and including Dr. Y. Takeda (NHW), Dr. C. Kumkumian (FDA) and representatives of the trade associations from Europe (EFPIA), USA (PMA) and Japan (JPMA).

It was discussed at the first international conference on Harmonization in Brussels and then again revised.

It is followed by the paper "general concept for stability testing", considering the draft of the harmonized guideline.

The last chapter consists of two papers. The first gives an excellent overview on the application of cyclodextrins, stability and their influence on the stability of active substances.

Finally the stability and stability testing of drug products derived by recombinant DNA technology are presented.

It is hoped that this book may help that stability testing can be carried out in the most efficient way for registration in the EC, Japan, the USA and worldwide.

B Contributors

A. C. Cartwright
Department of Health and Social Security/Medicines Division
Market Towers
1 Nine Elms Lane
GB-London SW8 5NQ

Dr. T. G. Beaumont
The Wellcome Foundation Ltd
Temple Hill
GB-Dartford/Kent

Dr. K. D. Gneuß
Dr. Karl Thomae GmbH
Birkendorfer Straße 65
D-W-7950 Biberach

Dr. W. Grimm
Dr. Karl Thomae GmbH
Birkendorfer Straße 65
D-W-7950 Biberach

Dr. H. Hoffmann
Dr. Karl Thomae GmbH
Birkendorfer Straße 65
D-W-7950 Biberach

Dr. K. Krummen
Sandoz AG
Analyt. Forschung und Entwicklung
Bau 360/1106
CH-4002 Basel

PD DR. Helga Moeller
Hoechst AG
Pharma Qualitätskontrolle, H 790
Postfach 80 03 20
D-W-6230 Frankfurt/M 80

Dr. C. Kumkumian
Food and Drug Administration
5600 Fishers Lane RM 14B-45
USA-Rockville, MD 20857

Dr. D. C. Monkhouse
Smith Kline Beecham Pharmaceuticals
Pharmaceutical Development
P. O. Box 1539
USA-King of Prussia, PA 19406-0939

Prof. Dr. F. Pellerin
Université Paris-Sud
Centre D'Etudes Pharmaceutiques
Laboratoire de Chemie Analytique
1. Avenue Jean-Baptiste-Clement
F-92296 Châtenay Malabry

Dr. P. H. Stahl
Ciba-Geigy AG
K-401.3.67
Klybeckstraße
CH-4002 Basel

Dr. W. Steuer
Boehringer Ingelheim GmbH
Binger Straße
D-W-6507 Ingelheim/Rhein

Dr. L. Szente
Cyclolab
Endrödi S. 38/40
H-1026 Budapest

Dr. Yasushi Takeda
National Institute of Hygienic Sciences
1-43 Hoenzaka 1-Chome,
Chuo-Ku, Osaka
540 Japan

Contents

C Lectures

1 Basic Principles of Stability Testing

Wolfgang Grimm, D-Biberach/Riß

1. Aim of Stability Testing

Stability testing is performed to ensure that drug products retain their full efficacy up to the end of their expiration date. Extensive analytical investigations are necessary for this purpose.

The results obtained are used to formulate the stability information, which ensures the quality, efficacy and safety of the drug product.

This means that all

- ☐ organoleptic,
- ☐ physico-chemical,
- ☐ chemical,
- ☐ microbial

test results relevant to the stability of the product must be within the predefined tolerance ranges until the end of the expiration date. The stability information is thus an important element in the quality assurance of a drug product.

1.1 Stability Information

In order that the quality of the drug product can be assured by the stability information, the following aspects must be taken into account:

☐ **Expiration date or shelf-life**

The expiration date must always refer to the climatic zone or zones in which the drug product is to be distributed. During the development of a drug product, a distinction is drawn between:

- expiration date for clinical samples

This is generally batch-specific;

- tentative expiration date

This is based on a prediction and has not yet been covered by the full storage period;

- final expiration date

Expiration date equivalent to storage period, final dosage form, final packaging material.

☐ **Storage instructions**

For certain drug products, the shelf-life can only be guaranteed if certain specific storage instructions are complied with.

☐ **Use life, stability in opened container**

For certain drug products it is necessary to specify a period within which the product must be used after the container has been opened.

☐ **Overage**

For certain drug products an overage is necessary in order to achieve an acceptable shelf-life. An overage may only be added if the formulation is already optimised.

1.2 Problems

Every physico-chemical system is constantly subject to changes which only reach a standstill at the absolute zero point. This natural law mandates the performance of stability testing. The following potential reactivities and interactions can be distinguished:

☐ reactivity of the active ingredient
☐ reactivity of the excipients
☐ reactivity of the packaging material
☐ interactions between excipients
☐ interactions between active ingredient and excipients
☐ interactions between active ingredient and packaging materials
☐ interactions between excipients and packaging materials
☐ interactions between active ingredient, excipients and packaging materials.

All reactions and the changes resulting therefrom can be influenced by:

☐ factors relating to manufacture
  - batch size
  - equipment
  - sequence in which constituents of the formulation are added
  - different quality of active ingredients, excipients, packaging materials, although they were all analysed before use and complied with the tolerances

☐ external factors
  - temperature
  - humidity
  - light
  - oxygen
  - pH

During the lifetime of a drug product from manufacture until use, the following external factors can influence its stability: [1]

☐ manufacture

- □ storage at the manufacturer
- □ transport
- □ storage at the wholesaler
- □ storage in the officinal or hospital pharmacy
- □ storage in the household medicine cabinet.

Stability testing therefore has to deal with a wide diversity of complex problems. The stability information, which covers all test parameters and is binding for all batches, has to be generated stepwise.

It is the outcome of all the stability tests that accompany the development of a drug product from the pilot tests on the active ingredient to the stage of continuous production.

When establishing a stability programme, the following points must be considered:

- □ active ingredients can be:
  - manufactured by chemical synthesis, produced by biochemical or biotechnological methods, or by gene technology,
  - obtained from plant or animal products.
- □ From this wide variety of active ingredients, a large number of dosage forms are developed.
- □ The drug products thus developed are to be distributed
  - in a country
  - in a climatic zone
  - in several climatic zones
  - worldwide.

Two completely different approaches are possible when implementing a stability programme:

- □ Formalistic stability programme

  The necessary investigations are defined in advance for each individual case. The advantage of this approach is that during development it is only necessary to work through the checklist provided, and the results can be ticked off correspondingly quickly by the examining authority.

  If the authority requests a foramalistic stability programme, it automatically assumes a major part of the responsibility at the same time.

  The following disadvantages should be mentioned:
  - The stability programme is very extensive, and thus clumsy, bureaucratic.
  - It requires constant revision in the light of new developments, although stability tests have to be designed to run over several years.
  - It fosters a formalistic approach in which the actual problems are easily overlooked.
  - The experience and expertise of the developer remain largely unused.
  - Harmonisation is difficult because of the formalistic differences.
- □ Problem-orientated stability programme based on principles.

  This type of stability programme is based on principles which are decisive for stability testing. Once they have been defined and described, and once they

are generally accepted, so too are the results which are obtained from a stability programme implemented along these lines. This also offers a real opportunity to achieve harmonisation.
A system of this kind can assimilate new scientific knowledge and developments; it is dynamic.
The stability programme for a development is attuned to the problems to be dealt with, and requires the active cooperation of the developer.

2. The Principles of Stability Testing

The principles will first be enumerated, then defined individually and described in detail.

☐ Systematic development
Stability testing accompanies the development of a drug product in the form of a logically structured system subdivided into individual stages.

Further principles are

☐ Selection of batches and samples
☐ Test criteria
☐ Analytical methods
☐ Tolerances
☐ Storage conditions
☐ Testing intervals
☐ Storage period
☐ Number of batches
☐ Packaging materials
☐ Evaluation
☐ Information

2.1 Systematic Stability Testing during the Development of a Drug Product

Comprehensive and reliable stability information is based on a large number of individual results generated during the development of a drug product. To allow the results to be summarised in the form of a general statement, however, the individual stages during development must be logically coordinated with each other.

The process of development can be divided into 6 steps: [2]

☐ Tests with the active ingredient
☐ Preformulation and formulation finding for the clinical samples and the final dosage form
☐ Stress and acceleration tests with selected formulations, selection of packaging materials

- ☐ Long-term testing
- ☐ Follow-up tests
- ☐ Tests in the event of changes during continuous production
  - major change.

Each of these six stages has a specific objective which determines its scope and level of effort. A stability programme frequently concentrates mainly on long-term testing. The required scope of testing is correspondingly extensive. In a general concept, long-term testing represents a step of six.

All the subsequent principles from the selection of batches up to the evaluation and assessment apply for each of these six stages. At the same time they also pass through a process of development, and have differing degrees of importance.

2.2 Selection of Batches and Samples

Instabilities can only be identified by comparing a recorded value with an initial or previous value. They can be regarded as definitely identified only when the analysis samples, the batches are uniform.

Basically the batches can be classified as follows:

- ☐ **Test or experimental batches**

  These are obtained during development and optimisation and are frequently small batches for which the process has not yet been validated. Strictly speaking, therefore, only the batch and not the product as a whole can be assessed on the basis of the test results. These batches are often clinical batches.
- ☐ **Representative batches**

  These batches are representative of the product in terms of composition, manufacture and packaging. As a rule, therefore, they have been produced by a validated process. The batch size is not the decisive point.

  A laboratory or pilot plant batch can be representative if they can be manufactured reproducibly for this batch size and stage of development.

  The results obtained with a representative batch can always be generalised, and allow the product to be assessed in its totality.

  Despite this strict differentiation between test and representative batch, in practice the results frequently exhibit no significant differences, especially for liquid and solid dosage forms.

Test and representative batches are however subject to three further requirements:

- ☐ The batches must be homogeneous and must consist of random samples.
- ☐ The containers put into storage must be representative of the batch.
- ☐ The samples investigated from a container must be representative of all the samples in the container at the time of the analysis.

2.3 Test Criteria

In stability testing, the criteria of a drug product are investigated
- ☐ which are potentially susceptible to change during the course of storage,
- ☐ which are especially important for quality, efficacy or acceptance.

The resulting test parameters can be divided into the following groups:
- ☐ organoleptic,
- ☐ physico-chemical,
- ☐ chemical properties, i.e. active ingredient assay and purity. This also includes determination of the decomposition products;
- ☐ microbial properties.

In some cases toxicological properties are also listed, but these are usually dealt with when determining and establishing limits for the decomposition products. The selection of specific test parameters for long-term testing is generally made on the basis of the results of the stress tests.

2.4 Analytical Methods

When the relevant test criteria have been established, suitable test methods are required for each test parameter.

These methods must ensure that any changes which may occur can be detected at the earliest possible stage and at the level of statistical significance.

The test or analytical methods must be validated, and the apparatus used must be qualified. The term "test or analytical methods" is understood to mean the entire analytical procedure. This embraces the following:
- ☐ reagents
- ☐ preparation of samples
- ☐ analytical method
- ☐ calibration
- ☐ evaluation
- ☐ calculation
- ☐ documentation

An important point in this context is the definition and specification for the reference standard.

The analytical validation should demonstrate that a test procedure is suitable for its intended purpose. For the assay determination in stability testing, this means that it is possible to determine whether the active ingredient content has changed in relation to its initial or previous value during the course of storage.

For the determination of assay and decomposition, the following points must be considered in the validation:
- ☐ specificity
- ☐ linearity of active ingredient and decomposition product
- ☐ accuracy
- ☐ precision

- ☐ limit of quantitation for the decomposition product
- ☐ ruggedness
- ☐ documentation

When the method is specific, the two most important points are:

- ☐ precision, the relative standard deviation RSD should be $< 1.5\%$.
- ☐ limit of quantitation, which should be at least 0.5% with reference to the active ingredient.

Possible impurities of the active ingredient are not followed up in stability testing. The analytical methods are optimised with reference to active ingredient and decomposition. If possible impurities are also potential decomposition products, they are naturally included in stability testing.

2.5 Tolerances

Relevant test criteria and specific methods suitable for stability testing make it possible to assess a sample. The basis for assessing its quality and for making decisions is provided by the specifications and tolerances. The specifications and tolerances must take account of three criteria in stability testing:

- ☐ manufacture,
- ☐ analysis,
- ☐ tolerable changes during storage.

In quality control, the tolerances for manufacture and analysis are applicable for freshly manufactured product. In stability testing, changes during storage also have to be taken into account. These tolerances must be established for all test parameters. They apply up to the end of the shelf-life.

The tolerance for tolerable changes during storage is decisively important for the quality and shelf-life and must be established with great care. [3]

The first step towards this is to ascertain the discriminatory power of the individual analytical procedures. For assay determiniations, this is done on the basis of the relative standard deviation. A similar approach can also be adopted for the physico-chemical test parameters hardness, disintegration, dissolution rate. In this case, however, the statistical data are batch-related.

When these data have been determined, statements can be made as to when a change has become statistically significant. Only then is it possible to establish tolerances for tolerable changes. It must be repeatedly emphasised that organoleptic, physico-chemical or microbial alterations have the same importance as a decrease of active ingredient content and decomposition.

2.6 Storage Conditions

Stability testing is concerned with detecting and assessing the changes that occur during storage and with formulating the stability information on the basis of the results.

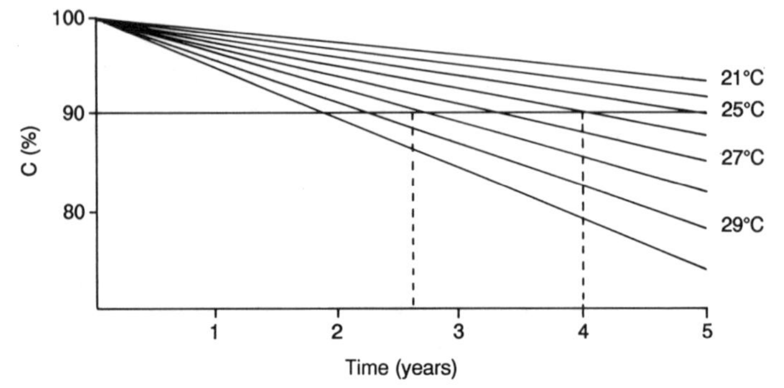

Fig. 1: Potency of active ingredient as a function of the temperature. Base value: 25 °C, 5 years, 90%

The extent of these changes depends to a critical degree on the storage conditions, temperature and relative humidity or water vapour partial pressure.

The influence of temperature can be described very accurately by applying the laws of reaction kinetics.

A model calculation of this type is presented in Figure 1 [4].

However, the influence of moisture is not taken into account here. The relationship between the vapour pressure of water and the absolute temperature is expressed by the Clausius-Clapeyron equation

$$\ln P = \frac{\Delta Hv}{R} \times \frac{1}{T} + \text{constant}$$

At a given temperature the partial pressure of water vapour in humid air cannot be greater than the vapour pressure of water of the same temperature, the saturation pressure P_s.

Table 1: Saturation pressure and absolute humidity in the temperatur range 0–50 °C

temperature (°C)	0	5	10	15	20	25	30	35	40	45	50
saturation pressure (mbar)	6.10	8.72	12.27	17.04	23.37	31.67	42.43	56.24	73.78	95.90	123.40
absolute humidity (g/m³)	4.84	6.80	9.40	12.82	17.29	23.04	30.36	39.60	51.14	65.44	82.98

Table 2: Dependence of changes in r.H. on temperature at a constant partial pressure of 6.1 mbar

temperature (°C)	0	5	10	15	20	25	30	35	40	45	50
r.H. (%)	100	70	50	36	26	19	14	11	8	6.4	4.9

Non-saturated air contains water vapour, the partial pressure P_D of which is lower than the saturation pressure p_s belonging to the respective temperature.

If the partial pressure P_D remains constant and if the temperature changes, P_s changes according to the Clausius-Clapeyron equation, and consequently so does the relative humidity.

Table 1 gives the saturation pressure and absolute humidity of water vapour for the temperature range 0 °C–50 °C.

Table 2 shows how the relative humidity decreases when the water vapour partial pressure remains the same, but the temperature increases. [5]

The storage conditions, i. e. temperature and humidity can lead to

- ☐ organoleptic,
- ☐ physico-chemical,
- ☐ chemical,
- ☐ microbial changes.

Especially with solid dosage forms, the sorption behaviour has to be considered by fixing storage conditions.

At higher temperatures, e. g. 40 °C desorption and loss of moisture can occur also at higher relative humidities as shown by figure 2 [5].

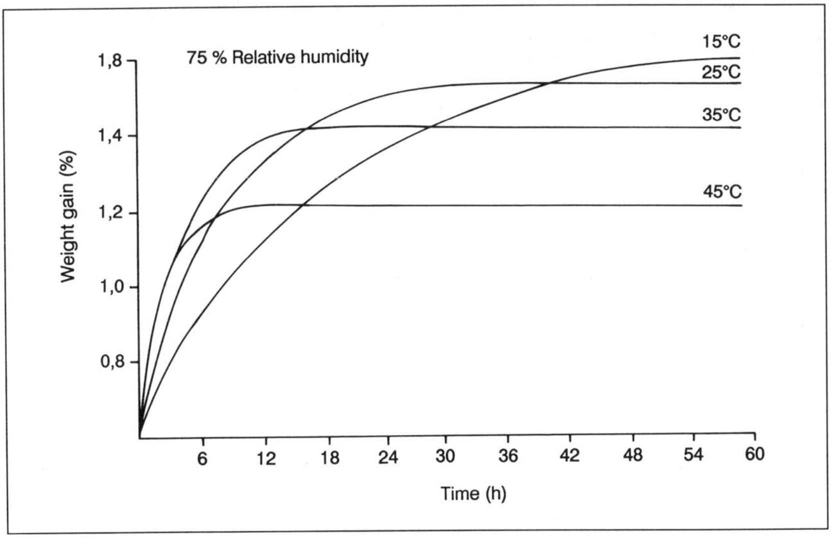

Fig. 2: Adsorption as a function of the temperature

When working out and establishing storage conditions, two decisive points must be remembered:

- ☐ clear differentiation between storage conditions for
  - stress and acceleration tests
  - long-term testing
  - follow-up stability testing
  - major-change stability testing
- ☐ logical relationship between the individual storage conditions, since only on this basis can a comprehensive general statement be made.

2.6.1 Storage conditions for stress and acceleration tests

Stress and acceleration tests are used to

- ☐ identify the weaknesses of an active ingredient or a formulation,
- ☐ make stability predictions,
- ☐ obtain information on extreme loading during transport.

The last two points require that a logical relationship should exist between the storage conditions for long-term tests and for stress and acceleration tests. Otherwise a stability prediction is not possible.

By increasing the temperature and applying the laws of reaction kinetics, stability predictions can be made for the chemical stability of the active ingredient in the drug product. The precondition for this is that ΔE, the activation energy, is not temperature-dependent, that the reaction mechanism does not change in the observed temperature range and for the extrapolated temperatures. The precondition for a change in the reaction mechanism would exist

- ☐ if the stress samples of solid dosage forms would lose moisture to different degrees at the individual temperatures,
- ☐ if the concentration of active ingredient or the ion strength increased to different degrees in semi-solid and liquid dosage forms.

This can only be prevented by the use of packaging materials which are not permeable to water vapour.

Predictions relating to organoleptic and physico-chemical properties cannot be made using the laws of reaction kinetics.

The organoleptic and physico-chemical changes observed at higher stress temperatures are recorded, but generally cannot be used to make predictions.

From this it is concluded that:

Stability predictions based on the results of stress and acceleration tests are only possible if a clear distinction is drawn between:

- ☐ storage conditions to test for chemical changes of the active ingredient in the drug product,
- ☐ storage conditions to test for organoleptic and physicochemical changes.

The storage temperatures used to test for chemical changes are within the range between 40 °C and 80 °C [6].

It thus becomes possible to accelerate the rate of decomposition to such an extent that relevant results are available within three months on the basis of which stability predictions can be made applying the laws of reaction kinetics.

The temperatures are a continuation of those used for long-term testing. The temperature is increased in 10 °C steps: (21 °C, 25 °C, 30 °C), 40 °C, 50 °C, 60 °C, 70 °C, 80 °C.

Which and how many of these are used depends on the problem under consideration and the dosage form. For a reliable stability prediction, however, at least three temperatures will be necessary.

The storage conditions for stress testing for organoleptic and physico-chemical changes differ depending on the dosage form:

Solid dosage forms:

The samples are stored in open containers until equilibrium is reached under the storage conditions of long-term testing and two further frequently used stress conditions:
21 °C/45%, 25 °C/60%, 25 °C/75%, 30 °C/70%, 40 °C/75%.

Semi-solid dosage forms:

≥ – 10 °C, temperature variable between 4 °C and 40 °C.

Liquid dosage forms:

≥ – 10 °C

The influence of humidity and temperature on chemical stability is tested by storing in parallel two samples of different initial moisture content in packaging materials impermeable to water vapour (e.g. glass container with twist-off closure) at temperatures between 40 °C and 80 °C.

2.6.2 Storage conditions for long-term testing

The binding expiration dates are established on the basis of the results of the long-term tests. The storage conditions necessary for these tests should correspond to the actual climatic conditions of the country or climatic zones in which the drug product is to be distributed.

Statements regarding the climatic conditions prevailing in the different locations can be made very precisely and reliably, since measured values for temperature and relative humidity or water vapour partial pressure representative of a period of up to 50 years are available for many cities.

If a stability study is to have validity for several countries or even worldwide, it is not possible to perform specific storage tests for each country.

Instead, a simplified approach is adopted whereby the world is divided into four climatic zones [8], to which the individual countries can then be assigned. The four climatic zones are defined as follows:

I temperate climate
II subtropical and Mediterranean climates
III hot, dry climate
IV hot, humid climate

The relevant storage conditions can now be established on the basis of the values measured for a number of cities in the relevant climatic zone[5].

The following two points should be taken into account in this process:

- □ The climatic data were measured in the open air, whereas drug products are stored in rooms. Consequently, 19 °C is used as the temperature for all months during which the temperatures measured are below 19 °C. The relative humidity is then converted to the correct level for this 19 °C by means of a suitable calculation.
- □ When establishing the mean value for the temperature, the formula of Haynes [8], by means of which the mean kinetic temperature can be calculated, is used especially in the presence of relatively marked temperature differences.

  For instance, if a sample is stored for one month at 20 °C and after one month at 40 °C, the resulting arithmetic mean is 30 °C.

  If however the temperature dependence is taken into account in accordance with the Arrhenius equation, from which Haynes derived his formula, the resulting kinetic mean is 34.2 °C. The higher temperature stress at 40 °C is taken into account accordingly.

The countries of today's most important markets for drug products, namely the EC, Japan and USA, are all classified as belonging to climatic zone II. It is therefore possible to establish uniform storage conditions corresponding to the same climatic conditions for these countries.

It is however important to note that the greater part of the USA and the EC are in fact located in climatic zone I; for Japan, only the Island of Hokkaido falls within this category.

The selection of the different cities for the four climatic zones has been done with the criteria and guide values of table 3 [5].

Table 3: Criteria and guide values for assignment of a city to the correct climatic zone

criteria	guide values for individual climatic zone I	II	III	IV
mean annual temperature measured in the open air	up to 15 °C	> 15–22 °C	> 22 °C	> 22 °C
calculated mean annual temperature (< 19 °C)	up to 20.5 °C	> 20.5–24 °C	> 24 °C	> 24 °C
mean annual water vapour partial pressure	up to 11 mbar	> 11–18 mbar	up to 15 mbar	> 15 mbar

Climatic zone I

Firstly, a list of cities of climatic zone I is compiled for the EC, Japan and USA, and the storage condition for climatic zone I is established. The measured data listed in each case originate from the following literature references [9–12].

Tables 4 and 5 list cities of climatic zone I in Europe and USA.

Table 6 presents a comparative review of the data. The climatic conditions exhibit a very good degree of correspondence.

The calculated data for climatic zone I are therefore:

20 °C, 42% r. H. equivalent to 9.9 mbar.

Room temperature for climatic zone I is defined as 15–25 °C. From this the mean kinetic temperature is calculated as 21.1 °C.

21 °C is therefore established as storage temperature. In order to have a safety margin for the partial pressure as well, this is established as 11.2 mbar equivalent to 45% r. H.

The storage condition established for climatic zone I is therefore:

21 °C 45% r. H. equivalent to 11.2 mbar

Table 4: Cities of climatic zone I in the EC

city	measured data		calculated data			4 hottest months			
	tempe-rature [°C]	partial pressure [mbar]	temp. 19 °C used [°C]	rel. humidity [%]	partial pressure [mbar]	[°C]			
Dublin	11.6	10.6	19.0	48	10.6	19.3	19.2	19.0	19.0
London	12.1	10.0	19.2	45	10.0	20.2	19.7	19.0	19.0
Copenhagen	10.0	9.5	19.2	43	9.5	20.4	19.4	19.0	19.0
Amsterdam	10.3	10.5	19.3	47	10.5	20.3	20.3	19.6	19.0
Berlin	11.0	10.0	19.6	44	10.0	21.4	21.2	20.7	19.3
Frankfurt	10.7	9.7	19.8	42	9.7	21.9	21.7	21.2	19.9
Munich	10.4	9.1	19.5	40	9.1	21.1	20.9	20.2	19.3
Paris	13.5	10.2	19.8	44	10.2	22.1	21.6	21.0	20.1
Lyons	13.9	10.8	20.1	46	10.8	23.4	23.1	21.6	20.8
Brussels	11.9	10.7	19.4	48	10.7	20.7	20.6	19.9	19.3
Luxembourg	8.8	9.7	19.4	43	9.7	20.9	20.6	20.1	19.0
Bordeaux	12.5	12.3	20.0	53	12.3	22.4	22.2	21.4	21.1
Marseilles	14.2	11.3	20.5	47	11.3	24.0	23.7	22.6	22.1
Strasbourg	10.0	10.4	19.7	45	10.4	21.9	21.6	21.0	19.9
Hamburg	8.7	9.3	19.3	55	9.3	20.7	20.5	19.8	19.0
Stuttgart	9.7	9.1	19.6	40	9.1	21.5	21.4	20.8	19.5
Dresden	9.2	9.7	19.6	43	9.7	21.5	21.4	20.8	19.5
Toulouse	12.7	12.1	20.1	51	12.1	22.9	22.9	21.7	21.4
Bolzano	12.0	10.7	20.4	45	10.7	23.6	23.2	22.5	21.8
mean	11.2	10.3	19.6	45	10.3	21.5	21.4	20.6	19.9

Table 5: Cities of climatic zone I in the USA

city	measured data		calculated data			4 hottest months			
	temp. [°C]	partial pressure [mbar]	temp. 19°C used [°C]	rel. humidity [%]	partial pressure [mbar]	[°C]			
New York	12.5	10.7	20.6	44	10.7	24.9	23.9	22.0	21.6
St. Louis	13.0	10.3	21.1	41	10.3	25.6	24.9	24.3	23.2
Chicago	10.5	9.9	20.4	41	9.9	24.3	23.5	22.7	21.4
Boston	10.8	9.9	20.1	42	9.9	23.4	22.9	21.8	21.0
Buffalo	8.2	9.0	20.0	39	9.0	22.9	22.5	21.5	20.5
Columbus Ohio	11.0	10.9	20.8	44	10.9	24.7	24.2	23.6	22.2
Denver	9.7	7.4	20.7	30	7.7	25.2	24.8	23.4	22.6
Philadelphia	11.9	10.5	20.6	43	10.5	24.5	23.9	23.3	22.1
Portland	7.2	8.5	19.8	37	8.5	22.7	22.4	21.0	20.1
Salt Lake City	10.5	6.8	21.2	27	6.8	26.8	25.9	23.9	23.0
San Francisco	13.8	11.5	19.1	52	11.5	19.8	19.6	19.0	19.0
Omaha	10.8	10.8	21.0	43	10.8	25.8	24.8	23.6	23.0
Seattle	10.6	9.7	19.6	43	9.7	21.6	21.4	20.0	19.9
Pittsburgh	10.2	9.6	20.4	40	9.6	23.8	23.4	22.8	21.6
Albuquerque	13.7	6.5	21.4	26	6.5	26.0	25.2	25.1	23.5
Detroit	10.1	9.2	20.3	39	9.2	23.9	23.4	22.6	21.2
Madison	7.3	9.1	20.1	39	9.1	23.5	22.9	22.0	20.5
Providence	10.1	9.3	20.1	40	9.3	23.2	22.9	21.8	21.0
mean	10.7	9.4	20.5	39	9.4	24.0	23.5	22.5	21.6

Table 6: Climatic zone I:
Summary of mean values for EC, Japan, USA

country	measured data			calculated data			4 hottest month			
	temp. [°C]	rel. humidity [%]	partial pressure [mbar]	temp. 19°C used [°C]	rel. humidity [%]	partial pressure [mbar]	[°C]			
EC $\bar{x}_{19}$	11.2	78	10.3	19.6	45	10.3	21.5	21.4	20.6	19.9
Japan $\bar{x}_1$	7.6	93	9.8	19.7	42	9.8	22.7	21.8	20.5	19.8
USA $\bar{x}_{18}$	10.7	74	9.4	20.5	39	9.4	24.0	23.5	22.5	21.6
$\bar{x}_{38}$	10.9	75	9.9	20.0	42	9.9	22.7	22.4	21.5	20.7

Table 7: Extreme daily temperatures during the 3 hot months and the hottest month, which are covered by the storage temperature 21 °C

temperature/time	8 h	6 h	4 h	6 h	MKT
23 °C/3 months	20 °C	22 °C	26 °C	24 °C	22.7 °C
25 °C/1 month	20 °C	24 °C	31 °C	26 °C	25.1 °C

The climatic values are likely to vary within the following limits:

21 °C ± 2 °C, 45% r. H. ± 5%

A symmetrical variation is assumed. In accordance with the calculated climatic data and taking into account the four hottest months, the following annual temperature pattern is obtained for climatic zone I:

19 °C/8 months, 23 °C/3 months, 25 °C/1 month equivalent to 20.8 °C as mean kinetic temperature (MKT).

This storage condition also takes daily temperature variations into account. See Table 7.

Climatic zone II

Measured and calculated data are first presented for one city each of the EC, Japan and USA, which is followed by a list for Europe, Japan and USA. See Tables 8, 9, 10, 11, 12.

The calculated data for climatic zone II are therefore:

21.6 °C, 52% r. H. equivalent to 13.5 mbar

Room temperature for climatic zone II is defined as 15 °C–30 °C. From this the mean kinetic temperature is calculated as 25,1 °C. Consequently, if 25 °C is established as the storage temperature for climatic zone II, this means that it is 3,4 °C above the calculated temperature of 21,6 °C.

The water vapour partial pressure of 13,5 mbar represents 43% r. H. at 25 °C. This would be to low corresponding to the temperature. It is therefore increased to 19 mbar. The storage condition etablished for climatic zone II is therefore:

25 °C, 60% r. H. equivalent to 19 mbar

The climatic values will probably vary symmetrically within the following limits:

25 °C ± 2 °C 60% r. H. ± 5%.

In accordance with the calculated climatic data and taking into consideration the four hot and two hottest months, approximately the following annual temperature pattern is obtained for climatic zone II:

Table 8: Measured and calculated data for three cities of climatic zone II

city	climate	January		February		March		April		May		June		July		August		September		October		November		December	
		7 a.m.	2 p.m.	7 a.m.	2 p.m.	7 a.m.	2 p.m.	7 a.m.	2 p.m.	7 a.m.	2 p.m.	7 a.m.	2 p.m.	7 a.m.	2 p.m.	7 a.m.	2 p.m.	7 a.m.	2 p.m.	7 a.m.	2 p.m.	7 a.m.	2 p.m.	7 a.m.	2 p.m.
Barcelona	measured (°C)	6.3	12.7	7.1	13.7	8.9	15.6	11.0	18.2	14.0	21.4	18.1	25.4	20.7	27.9	20.8	27.7	18.7	25.0	14.5	20.5	10.5	16.4	7.5	13.1
	(%)	74	61	71	58	75	60	73	59	72	59	68	59	70	59	75	63	79	66	77	64	75	64	72	62
	19°C used	19	19	19	19	19	19	19	19	19		19						19		19		19	19	19	19
	calculated (°C/%)	19/37		19/37		19/44		19/50		20.2/56		22.2/62		24.3/65		24.3/69		22.0/72		19.8/61		19/49		19/38	
Tokyo	measured (°C)	−1.4	8.7	−0.5	9.0	2.9	12.4	3.8	17.8	13.2	22.0	17.3	24.8	21.5	28.7	22.6	30.2	19.1	26.5	13.0	20.4	6.8	15.7	1.2	11.0
	(%)	73	48	71	48	75	53	81	59	85	62	89	68	91	69	92	66	91	68	88	64	83	58	77	51
	19°C used (%)	19	19	19	19	19	19	19	19	19		19								19		19	19	19	19
	calculated (°C/%)	19/20		19/23		19/32		19/43		20.5/62		21.9/74		25.1/82		26.4/80		22.8/79		19.7/63		19/43		19/27	
Atlanta	measured (°C)	2.9	11.1	3.6	12.1	5.8	15.7	10.1	21.2	15.1	26.1	19.7	29.8	21.5	30.6	21.0	30.3	17.9	27.7	11.3	22.4	5.3	16.1	2.8	11.3
	(%)	79	59	76	55	77	50	80	52	83	54	86	60	90	64	91	62	89	60	84	53	82	54	80	60
	19°C used	19	19	19	19	19	19	19		19								19		19		19	19	19	19
	calculated (°C/%)	19/32		19/31		19/36		20.1/49		22.6/59		24.8/71		26.1/75		25.7/75		23.4/71		20.7/52		19/39		19/32	

Table 9: Cities of climatic zone II in the EC

city	measured data		calculated data			4 hottest months			
	tempe-rature	partial pressure	temp. 19°C used	rel. humidity	partial pressure				
	[°C]	[mbar]	[°C]	[%]	[mbar]	[°C]			
Barcelona	16.5	14.4	20.6	59	14.4	24.3	24.3	22.2	22.2
Madrid	13.9	10.1	20.6	42	10.1	24.9	24.3	23.0	22.0
Malaga	18.5	15.2	21.2	60	15.2	25.4	25.3	23.5	22.8
Lisbon	16.6	13.5	20.6	56	13.5	23.4	23.2	22.5	21.9
Athens	18.3	12.5	21.9	48	12.5	28.0	28.0	25.0	24.2
Rome	16.2	14.8	21.1	49	12.4	24.7	24.7	23.2	22.7
Naples	16.4	13.7	20.9	56	13.7	25.2	25.0	22.7	22.5
Palermo	18.2	13.2	21.2	52	13.2	24.8	24.7	23.7	23.2
Saloniki	16.1	13.0	21.4	51	13.0	26.7	26.4	24.2	23.4
Cadiz	18.1	15.6	20.8	64	15.6	24.5	23.8	23.3	22.1
Seville	18.8	11.6	21.3	58	14.6	28.0	27.9	25.6	25.5
Heraklion	18.6	14.2	21.3	56	14.2	25.6	25.4	23.2	23.1
Messina	18.0	15.0	21.2	60	15.0	26.3	25.8	23.6	23.0
Gibraltar	18.5	15.2	20.8	62	15.2	24.7	24.1	22.9	22.2
$\bar{x}$	17.3	13.8	21.1	55	13.8	25.5	25.2	23.5	22.9

Table 10: Cities of climatic zone II in Japan

city	measured data		calculated data			4 hottest months			
	tempe-rature	partial pressure	temp. 19°C used	rel. humidity	partial pressure				
	[°C]	[mbar]	[°C]	[%]	[mbar]	[°C]			
Hiroshima	14.8	13.7	21.0	55	13.7	26.6	25.4	22.9	22.1
Nagasaki	16.5	15.0	21.3	59	15.0	27.5	26.3	23.9	22.1
Osaka	15.5	14.3	21.4	56	14.3	27.9	26.7	23.7	22.8
Tokyo	14.6	13.8	20.9	56	13.8	26.4	25.1	22.8	22.1
$\bar{x}$	15.3	14.2	21.2	56	14.2	27.1	25.9	23.3	22.2

Table 11: Cities of climatic zone II in the USA

City	measured data		calculated data			4 hottest months			
	temp.	partial pressure	temp. 19 °C used	rel. humidity	partial pressure				
	[°C]	[mbar]	[°C]	[%]	[mbar]	[°C]			
Los Angeles	18.0	13.0	21.2	52	13.0	23.8	23.8	23.5	22.1
San Diego	17.4	14.2	20.5	59	14.2	22.3	22.2	22.0	21.3
Houston	21.1	18.3	23.3	64	18.3	29.0	28.7	27.7	26.7
Nashville	15.6	12.3	21.3	49	12.3	26.8	26.2	25.2	24.1
Atlanta	16.3	14.3	21.5	56	14.3	26.1	25.7	24.8	23.4
Birmingham	17.8	16.0	22.5	59	16.0	27.8	27.3	26.4	25.3
Charleston	18.5	16.3	22.3	61	16.3	27.0	26.5	26.2	24.2
Kansas City	13.7	12.2	21.5	48	12.2	27.5	26.5	24.4	23.6
Little Rock	16.5	14.7	22.2	55	14.7	27.8	27.4	26.0	24.6
Oklahoma	15.8	13.6	22.1	41	13.6	28.2	28.1	25.6	24.1
El Paso	17.3	7.8	22.7	28	7.8	27.7	27.1	26.9	25.0
Las Vegas	18.7	6.0	23.5	21	6.0	32.0	30.7	28.3	26.8
Phoenix	20.6	8.8	24.6	29	8.8	32.1	30.9	28.8	28.3
Asheville	12.6	12.0	20.7	49	12.0	23.8	23.6	23.2	22.2
New Orleans	21.3	19.7	23.2	69	19.7	28.6	28.4	27.8	26.9
Sacramento	15.7	10.9	21.9	41	10.9	26.7	26.1	25.1	24.7
Washington	13.9	11.2	21.0	45	11.2	25.7	24.7	23.8	22.2
$\bar{x}_{17}$	17.1	13.0	22.1	49	13.0	27.7	26.7	25.5	24.5

Table 12: Climatic zone II: Summary of mean values for EC, Japan, USA

country	measured data			calculated data			4 hottest month			
	temp.	rel. humidity	partial pressure	temp. 19 °C used	rel. humidity	partial pressure				
	[°C]	[%]	[mbar]	[°C]	[%]	[mbar]	[°C]			
EC $\bar{x}_{14}$	17.3	70	13.8	21.1	55	13.8	25.5	25.2	23.5	22.9
Japan $\bar{x}_4$	15.3	82	14.2	21.1	56	14.2	27.1	25.9	23.3	22.2
USA $\bar{x}_{17}$	17.1	64	13.0	22.1	49	13.0	27.7	26.7	25.5	24.5
$\bar{x}_{35}$	17.0	70	13.5	21.6	52	13.5	26.4	26.0	24.5	23.6

Table 13: Extreme daily temperatures during the 4 hot and the 2 hottest months of climatic zone II, which are covered by the storage temperature 25 °C

temperature/time	8 h	6 h	4 h	6 h	MKT
26 °C/3 months	21 °C	25 °C	31 °C	27 °C	26.1 °C
30 °C/1 month	24 °C	28 °C	36 °C	32 °C	29.9 °C

21 °C/6 months, 26 °C/4 months, 30 °C/2 months equivalent to 25.1 °C MKT.

If the temperature pattern during one day is divided up as follows:
8 hours night, 6 hours morning, 4 hours midday, 6 hours afternoon
the following picture emerges for the four hot and two hottest months:

The standard storage condition:

25 °C 60% r. H.

is therefore obtained for the EC, Japan and USA.

This storage condition meets the following requirements:

- ☐ It corresponds to the mean kinetic temperature of climatic zone II.
- ☐ It covers seasonal and daily temperature variations.
- ☐ For the temperature and water vapour partial pressure it is above the data calculated from the values measured for 35 cities.
- ☐ It has been derived logically and is scientifically based.

Climatic zone III, hot, dry

Based on the value 27.9 °C for the mean kinetic temperature and a safety margin of 2 °C, a storage temperature of 30 °C is obtained. A water vapour partial pressure of 15 mbar is established in order to maintain a relative humidity of 35%. The storage condition established for climatic zone III is therefore:

Table 14: Citys of climatic zone III

country	measured data		calculated data				4 hottest month			
	temp. [°C]	partial pressure [mbar]	temp. 19°C used [°C]	T_{MKT}* [°C]	rel. humidity [%]	partial pressure [mbar]	[°C]			
Aswan	27.0	10.0	28.3	30.3	25.4	10.0	33.7	33.4	32.1	31.2
Baghdad	22.7	11.0	25.7	27.0	33.6	11.0	34.5	34.0	32.2	30.4
Elat	25.0	12.0	26.4	28.3	34.9	12.0	33.4	33.0	31.1	31.1
Khartoum	29.5	11.8	29.9	32.5	28.0	11.8	33.8	33.5	31.8	31.3
Mossul	20.2	12.0	24.5	25.7	39.0	12.0	33.2	32.4	29.6	28.9
New Delhi	24.8	14.5	26.8	27.9	41.0	14.5	33.6	33.3	31.2	27.9
Riyadh	24.9	8.0	26.7	28.0	22.8	8.0	34.3	33.9	33.1	31.8
Suez	22.5	13.5	24.4	25.4	44.2	13.5	33.0	33.0	31.4	30.4
Jerusalem	22.8	14.5	24.6	25.8	46.9	14.5	29.7	29.1	28.3	27.8
$\bar{x}_g$	24.4	11.9	26.4	27.9	25.0	11.9	33.2	32.8	31.2	30.1

*T_{MKT}: The mean kinetic temperature is also listed here

30 °C 35% r.H. equivalent to 15 mbar

The climatic values are likely to vary symmetrically within the following limits:

30 °C ± 2 °C 35% ± 5% r.H.

In accordance with the calculated climatic data and taking the four hottest months into account, the following annual temperature pattern is obtained for climatic zone III:

25 °C/4 months, 30 °C/4 months, 34 °C/4 months equivalent to 30.3 °C MKT.

If the temperature course during one day is again divided up, the following maximum permissible temperatures are obtained for climatic zone III:

Table 15: Extreme daily temperatures during the 4 hot and the 4 hottest months of climatic zone III, which are covered by the storage temperature 30 °C

temperature/time	8 h	6 h	4 h	6 h	MKT
30 °C/4 months	24 °C	28 °C	36 °C	32 °C	29.9 °C
34 °C/4 months	25 °C	32 °C	42 °C	36 °C	34.2 °C

Table 16: Citys of climatic zone IV

country	measured data		calculated data				4 hottest months			
	tempe-rature [°C]	partial pressure [mbar]	temperature [°C]	T_{MKT^*}	rel. humidity [%]	partial pressure [mbar]	[°C]			
Bangkok	28.1	28.6	28.1	29.1	75.1	28.6	30.3	29.8	29.2	29.0
Belem	26.6	32.2	26.6	27.6	92.4	32.2	27.0	26.8	26.8	26.5
Bombay	26.9	27.0	26.9	27.6	76.0	27.0	29.7	28.7	28.1	26.2
Darwin	23.1	24.0	28.1	29.1	63.1	24.0	30.0	29.5	29.5	28.9
Jakarta	26.8	27.9	26.8	27.4	80.0	27.9	27.3	27.3	27.0	27.0
Calcutta	26.3	25.5	26.3	27.0	74.0	25.5	30.4	30.1	29.7	28.8
Manila	27.2	28.0	27.2	28.2	77.3	28.0	28.9	28.4	28.4	27.5
Maracaibo	29.0	28.5	29.7	31.0	73.0	28.5	29.8	29.8	29.5	29.5
Recife	26.3	15.8	26.3	26.7	75.3	28.5	27.5	27.5	27.3	27.3
Rio de Janeiro	23.0	22.0	23.3	23.8	77.0	22.0	26.3	26.0	25.3	24.8
Singapore	27.3	28.2	27.3	27.9	77.7	28.2	27.8	27.7	27.6	27.4
Taipei	22.1	21.0	23.4	23.8	73.0	21.0	28.7	28.4	27.8	26.8
$\bar{x}_{12}$	26.5	26.6	26.7	27.4	76	26.6	28.6	28.3	28.0	27.5

Climatic IV, hot, humid

Based on the value 27.4 °C for the mean kinetic temperature, a safety margin and the 30 °C for climatic zone III, a storage temperature of 30 °C is also established for climatic zone IV.

A water vapour partial pressure of 30 mbar equivalent to 70% r.H. is established.

The storage condition established for climatic zone IV is therefore

30 °C 70% r.H. equivalent to 30 mbar

The climatic values are likely to vary symmetrically within the following limits:

30 °C ± 2 °C 70% ± 5% r.H.

In accordance with the calculated climatic data, the following annual temperature pattern is obtained for climatic zone IV which is covered by the storage temperature of 30 °C.

27 °C/4 months, 29 °C/4 months, 31 °C/4 months equivalent to 29.1 °C MKT.

The daily variations are covered in the same way, and are as follows:

To summarise: the measured, calculated and derived data and the data measured in storage rooms [13] for the four climatic zones are listed comparatively in the table below.

The data derived for the four climatic zones fulfil the following criteria:

- ☐ They are obtained by means of a logically structured, clearly functioning system which takes account of temperature and humidity.
- ☐ They are based on measured climatic values.
- ☐ There is good correspondence between calculated data and data measured in storage rooms.
- ☐ The derived data include a safety margin for temperature and partial pressure in relation to the calculated data.
- ☐ For climatic zones I and II the derived storage temperatures correspond to

Table 17: Extreme daily temperatures during the 4 hot and the 4 hottest months of climatic zone IV, which are covered by the storage temperature 30 °C

temperature/time	8 h	6 h	4 h	6 h	MKT
29 °C/4 months	26 °C	28 °C	32 °C	30 °C	28.7 °C
31 °C/4 months	26 °C	28 °C	35 °C	32 °C	30.1 °C

Table 18: Comparative review for all four climatic zones

climatic zone	measured data			calculated data			derived data			data measured in storage rooms		
	[°C]	[%]	[mbar]	[°C]	[%]	[mbar]	[°C]	[%]	[mbar]	[°C]	[%]	[mbar]
I	10.9	75	9.9	20.0	42	9.9	21.0	45	11.2	18.7	45	9.8
II	17.0	70	13.5	21.6	52	13.5	25.0	60	19.0	21.1	52	13.1
III	24.4	39	11.9	26.4	35	11.9	30.0	35	15.0	26.0	54	18.0
VI	26.5	77	26.6	26.7	76	26.6	30.0	70	30.0	28.4	70	27.0

the respective mean kinetic temperature, calculated from the limit values for the definition of room temperature.

- ☐ Seasonal and daily variations in temperature are taken into account and contained in the respective storage temperature.
- ☐ The storage temperatures for climatic zones III and IV represent at 30 °C the upper limit value for the definition of RT in climatic zone II.

To allow a city to be assigned rapidly to the appropriate climatic zone, Table 3 provides guide values for temperature and pressure [15].

2.6.3 Storage conditions for ongoing and follow-up stability testing

The storage conditions for ongoing and follow-up stability testing largely correspond to those for long term testing.

2.6.4 Storage conditions for major-change stability testing

If a statement is to be achieved rapidly, the samples are stored in a standard packaging material at 40 °C besides the conditions of long term testing.

2.7 Testing intervals

A change can be identified and recorded as definite if the stored samples are investigated in time sequence. In addition to the time pattern, the temperature pattern is of great importance, especially in stress and acceleration tests.

When establishing the testing intervals, the time pattern, the following points must be taken into account:

- ☐ development status of the drug product
  - stress and acceleration tests
  - experimental batches

Table 19: Relationship between the number of analyses and the coefficient of variation

n	RSD 0.1	0.2	0.3	0.5	1.0	1.5	2.0	2.5	5.0
2	0.90	1.80	2.70	4.5	9.0	13.5	18.0	22.5	44.9
3	0.25	0.50	0.74	1.2	2.5	3.7	5.0	6.2	12.4
4	0.16	0.32	0.48	0.8	1.6	2.4	3.2	4.0	8.0
5	0.12	0.25	0.37	0.6	1.2	1.9	2.5	3.1	6.2
10	0.07	0.14	0.21	0.4	0.7	1.1	1.4	1.8	3.6

Table 20: Active ingredient assay [%]

storage condition	storage time [months] 3	6	9	12	18	24	36	48	60
21 °C/45%	99.5	99.0	98.5	98.0	97.1	96.1			
25 °C/60%	99.2	98.5	97.7	96.9	95.5	94.0			
30 °C/70%	98.7	97.4	96.1	94.9	92.4	90.0			
21 °C/45%	99.9	99.8	99.7	99.6	99.4	99.2	98.8	98.5	98.1
25 °C/60%	99.8	99.7	99.5	99.4	99.1	98.8	98.2	97.6	97.0
30 °C/70%	99.7	99.5	99.2	99.0	98.5	98.0	97.0	96.0	95.0

- representative laboratory and pilot plant batches
- representative production batches

☐ discriminatory power of analytical methods

☐ anticipated rates of decomposition

For which two examples:

Table 19 indicates the relationship between the number of analyses and the coefficient of variation (relative standard deviation) by means of which a significant deviation from the theoretical or initial value is identified [6].

The values summarised in Table 20 show rates of decomposition depending upon the storage period for an unstable and a stable product.

These two tables show very clearly that too short a time interval between tests is not justified particularly for stable products, because the decreases in the assay values cannot be detected analytically.

2.8 Storage Period

The storage period depends on the development status, the stability behaviour and the target expiration date.

In stress and acceleration tests the storage period should not exceed 3 months in order to ensure speedy development.

In long-term testing, an expiration date of not more than 5 years is generally regarded as optimum.

Under conditions that directly correspond to the climatic zones, storage should therefore be for 5 years, unless the tolerance limit is reached beforehand due to an instability.

2.9 Number of Batches

The number of batches must again be appropriate to the development status and the problems being examined.

2.9.1 New drug product, active ingredient, dosage form

- ☐ Stress and acceleration tests for active ingredient and formulation: 1 batch
- ☐ Long-term testing of active ingredient: 3 batches.
- ☐ Long-term testing of formulation: 3 batches.

The development of a drug product can be regarded as complete when the production becomes routine. The end of the first year of production can be regarded as the point of reference. During this phase of development, three representative batches are put into storage and tested.

If an influence of the batch size on stability is observed during scaling-up, three production batches must be included in stability testing.

2.9.2 Known active ingredients, new dosage form

The tests with the active ingredient are not necessary. Otherwise the same number of batches as used for Section 2.9.1.

2.9.3 New dosage

If existing results can be used, two representative batches are sufficient, one of them a production batch.

2.9.4 Major change

If a comparative acceleration test shows that the stability information obtained so far is not adversely influenced by the major change, two representative production batches are sufficient.

2.9.5 Follow-up

One batch per product and year, in the case of several dosages the most sensitive.

2.10 Packaging Materials

The primary packaging material is a constituent of the finished drug product and must therefore be included in the stability programme from the start of development onwards. The following points must be clarified/investigated in this process:

- ☐ statutory requirements, toxicological information
- ☐ exact description
  - type
  - additives
- ☐ physical properties
- ☐ general, generalisable tests
- ☐ interaction tests
- ☐ general suitability, specific suitability.

The effort required in each case depends on the type of packaging material and especially on the dosage form.

2.10.1 Solid dosage forms

If the results of the stress and acceleration tests for the bulk-ware and the information on the physical properties of potential packaging materials are available, the suitability of individual packaging materials can be established with a high level of certainty in many cases.

If, for instance, the samples are not sensitive to moisture, suitable tests in blisters are performed and all packaging materials that are less permeable to water vapour are suitable without the need to perform further stability tests.

Therefore, it is also not necessary to include different pack sizes in the stability programme. The same applies in the event of a major change of the packaging material.

If however the samples are sensitive to moisture, for example, the suitability of a more permeable packaging material or a new closure system must always be clarified by means of stability tests.

2.10.2 Liquid dosage forms

The most critical are infusion solutions and injectables in plastics, especially when elastomers are used. Interaction tests are always necessary for these forms, and the same applies in the event of a major change.

If different container sizes are used, it is not necessary to perform a new or parallel study in every case when the same plastic or elastomer is used.

2.10.3 Semi-solid dosage forms

Similar requirements apply as for the liquid dosage forms. However, it is only exceptionally necessary to test different tube sizes, and testing is confined to the "most sensitive". This applies especially in the case of metal tubes which are impermeable.

2.11 Evaluation

2.11.1 Critical evaluation of investigational results

The investigational results for all test criteria must be included in the assessment, not only those for the active ingredient assay and decomposition. The following stepwise approach is adopted:

- □ Batch in one packaging material
  - Investigations are performed for each individual test parameter to establish whether a significant change has occurred in relation to the initial value during the course of storage, and whether the tolerance limit was reached or exceeded. The laws of statistics are used for this purpose.
  - The results for the different storage conditions are compared.
- □ Batch in parallel packaging materials
  After evaluating the results of each batch in its packaging material, a comparison of packaging materials is performed.
- □ Comparison of batches
  - Experimental, representative batch
  - Representative batches of different size (scaling-up). In this case too, the laws of statistics will be applied as far as possible to achieve a comparative assessment.

2.11.2 Stability predictions

After evaluating the individual batches and performing the comparison of packaging materials and batches, a stability prediction is established on the basis of the results for as long as the expiration date and the storage period for which results are available do not yet correspond.

The results of the stress and the acceleration tests provide the basis for this prediction.

Here again, an extrapolation is performed for the stability behaviour of the active ingredient according to the laws of reaction kinetics [14].

At the same time, however, the results of the acceleration tests are used to establish organoleptic and physico-chemical criteria for the stability prediction.

In the same way, the results of long-term testing are used for the stability prediction.

2.12 Stability Information

2.12.1 Scope

The stability information applies for the product in its entirety, i. e. it is binding for each individual production batch. Consequently it is also based on the results obtained during the systematic development of the drug product.

The main points of emphasis are:

- ☐ active ingredient stability profile,
- ☐ results of the stress and acceleration tests,
- ☐ long-term tests.

The following points must be contained in the stability information, as far as necessary:

- ☐ expiration date for the countries and climatic zones in which the drug product is to be distributed,
- ☐ storage instructions,
- ☐ a use life,
- ☐ overage.

Expiration date or shelf-life

A formal distinction is made between the tentative and final expiration date. An expiration date is regarded as tentative as long as it is not covered by the full storage period. A period of 5 years is regarded as optimum, and, on the basis of a prediction, it is extended stepwise during the course of development.

The shelf-life is proposed by the manufacturer and either confirmed or modified by the regulatory agency.

Storage instructions

These may be necessary in order to ensure compliance with the expiration date and safeguard the stability of a drug product. They must be clear and as far as possible internationally standardised.

They are derived from the results of the stress and long-term tests with the following storage conditions:

≥ – 10 °C, 4 °C, 21 °C/45%, 25 °C/60%, 30 °C/70%, 40 °C

- ☐ Store up to 30 °C.
- ☐ Store up to 25 °C.
- ☐ Store up to 21 °C.
- ☐ Store at 2−8 °C refrigerate, do not freeze.
- ☐ Store below 8 °C.
- ☐ Store bewow −18 °C, deep freeze.
- ☐ Protect from light.

2.12.2 Period covered by stability information

☐ Clinical trial, CTX, IND documents
A minimum shelf-life of 6 months must be demonstrable. This is possible on the basis of:
- active ingredient stability profile
- results of stress and acceleration tests
- long-term testing, 1 batch, 3-month value

☐ Application for licence or registration
A minimum shelf-life of 24 months must be demonstrable. This is possible on the basis of:
- active ingredient stability profile
- stability prediction based on stress tests
- long-term testing, 3 representative batches, 6-month value.

Since the product is not yet on the market, there is no risk at this point.
The long-term tests are naturally continued parallel to the licensing procedure. If instabilities deviating from the prediction occur during this period, the authority is informed immediately.

☐ Notice of modification in the event of a major change
It must be ensured that the stability information obtained so far is not adversely influenced by a major change. This is possible by means of:
- comparative stress and long-term tests of the prechange status versus the major change, 3- and 6-month value.

References

[1] Grimm, W., in: Stability Testing of Drug Products. Wissenschaftliche Verlagsgesellschaft mbH, Stuttgart 1987
[2] PMA's Joint-PDS Stability Committee "Stability Concepts" Pharmaceutical Technology, 1984
[3] Grimm, W., Pharm. Ind. **45**, 715−720, 796−800, 1984
[4] Grimm, W., in: Oelschläger, H.: Fortschritte in der Arzneimittelforschung, Wissenschaftliche Verlagsgesellschaft mbH, Stuttgart 1984
[5] Grimm, W., Drugs made in Germany **29**, 1986
[6] Grimm, W., Schepky, G., Stabilitätsprüfung in der Pharmazie, Theorie und Praxis. Editio Cantor Verlag 1980
[7] Futscher, N., Schumacher, P., Pharm. Ind. **34**, 479−483, 1972
[8] Haynes, J.D., J. Pharm. Sci. **60**, 927−929, 1971
[9] Länder und Klima, Europa/UdSSR, F.A. Brockhaus, Wiesbaden 1982
[10] Länder und Klima, Nord- und Südamerika, F.A. Brockhaus, Wiesbaden 1982
[11] Länder und Klima, Asien/Australien, F.A. Brockhaus, Wiesbaden 1982
[12] Länder und Klima, Afrika, F.A. Brockhaus, Wiesbaden 1982
[13] Spingler, E., Verpackungs-Rundschau **3**, 17−21 (1974)
[14] Gneuß, K.D., Acta Pharm. Technol. **35**, 67−74 (1989)

2 Characterization and Improvement of the Stability Behaviour of Drug Substances

P. H. Stahl, CH-Basel

1. Introduction

The information by which a new chemical entity (= NCE; new active substance = NAS) is accompanied on the step from the research to the development stage, as a rule, is of merely limited value to the development pharmacist. The structural formula, melting point, and a few qualitative solubility data would constitute only a narrow information base to begin dosage form development. The chemist characterizes the new substance as a chemical structural entity whereas the efforts of the development pharmacist focus on the dosage form or therapeutic system to be designed for the new substance. The information gap between chemical research and pharmaceutical development is bridged by the Preclinical Preformulation Phase during which the new substance is characterized and probed as a physical and physicochemical entity from biopharmaceutical and galenical angles of view. This is done taking into account the possible indications and therapeutical exspectations as well as the envisaged dosage forms, the available technologies and the border conditions dictated by legal restrictions and environmental responsibility.

The objective of the preformulatory characterization is to establish a profile of a new active substance in order to be able to

- adequately process, handle and store the NAS,
- recognize weaknesses of the NAS,
- identify possible pitfalls during development,
- anticipate problems,

and, in consequence, adequately design and develop the intended dosage forms for the NAS. In this context stability considerations play a leading role.

Stability and stabilization is a topic leaning on a huge rock of experience and literature. Because it is virtually impossible to compress this into a short contribution, a rough outline is offered instead, including a few rock fragments and pebbles on techniques and own observations.

2. Preformulation characterization of an active substance: Stability screening in the preformulation program

Preformulation work sets out by looking at the structural formula of the NCE and trying to squeeze out from it as much information as possible. Thanks to the work of several groups of chemical physicists and physical pharmacists in academia and industry who succeeded in harnessing theory and observation for practical use in many cases, the quantitative prediction of key properties from chemical structure on the basis of empirical correlations and extrathermodynamic linear relationships is possible to a reasonably useful, sometimes even high degree of precision. This applies to the octanol/water partition coefficient [1], solubility [2] and ionization constants [3]. Thus, in this respect we don't have to start from zero with an NCE. However, when we ask for a similar predictive system for the stability of substances we are lost. Although – ironically – stability-related reactions like ester hydrolysis and esterification of benzoic acids were among the first ones investigated by Hammet to derive one of the key elements of extrathermodynamic relations, i.e. the substituent constant useful to predict the reactivity of further members of a series of compounds, we are still left with qualitative guesses or – at best – grossly ranking and relating assumptions about the stability and degradation behaviour of an NCE. The reasons are manifold: There is a multiplicity of possible reactions and interactions, and there are too many and to a vast extent hardly controllable and measurable factors which simultaneously act on a substance in a pharmaceutical product during its life cycle. In addition, the more phases constitute the system under consideration, e.g., a tablet, the less it is accessible for modelling and computation of reaction rates.

Any formulation development work requires information on properties of the NAS in its solid state, in the dissolved state and on interactions with other materials. However, in an industrial environment preformulation work oftentimes has to be performed under a series of restricting conditions:

- limited supply of sample material;
- limited time frame;
- limited resources;
- hence limitations to chose from the menu of possibilities.

Therefore, in the outset, the characterization of a NAS cannot be as complete as one might ideally wish but rather consists of a selection of the most important parameters.

Such a basic set or core-set of indispenable knowledge on fundamental properties is shown in Table 1. A few parameters must be known for any dosage form to be developed, the most prominent ones being solubility and stability data. Other ones differ in their bearing on the development of the various dosage forms. Around the core-set, dosage-form specific shells of additional knowledge have to be built later on.

Table 1: Significance of Properties for Dosage Form Development

Property, Parameter		Solid p.o.	Liquid p.o.	Par-entrl.	Top-ical	Trans-derm.	Inhal.-Aeros.
Solubility in water	pH=1...10	+	+	+	+	+	+
Electrolytic state	pKa	+	+	+	+	+	+
Lipophilicity	log $P_{Oct.}$	+	+	+	+	+	+
Stability	solid	+	+	+	+	+	+
	solution	+	+	+	+	+	+
Hygroscopicity		+	+/−	+/−	+/−	+/−	+
Morphology		+	+/−	−/+	+	+	+
Particle Size		+	+/−	−/+	+	+	+
Milling properties		+	+/−	−	+/−	+/−	+
Wettability, aqueous		+	+/−	+/−	+/−	+/−	+/−
Taste		+	+	−	−	−	+
Odour		+	+	−	+	+	+
Compatibility							
− tablet excipients		+	−	−	−	−	−
− other excipients		−	+	+	+	+	+
− container materials:							
− − glass		−	+	+	−	−	−
− − other		−	+/−	+/−	+	+	+
− corrosiveness		+	−/+	+/−	+/−	+/−	+/−
− other parenterals		−	+	−	−	−	−

\+ fundamental
− not required, nonessential
+/− useful; usually required, but under certain circumstances not significant
−/+ usually not important, but in certain cases required
(shaded) properties of fundamental importance

Stability aspects and stability testing comprise three capital groups: Physical, chemical and microbiological stability. As a special topic, the latter will not be considered here.

2.1 Physical Stability

Physical stability is concerned with the variability of the solid state of a substance (Table 2). More often than not, polymorphism and/or pseudopolymorphism is encountered in active substances. Polymorphs and pseudopolymorphs, i.e. solvates, and among them prominently the hydrates, differ in several pharmaceutically relevant properties. As a consequence of changing morphology of a solid, the properties change accordingly. In this case we have to deal with a type of physical instability although the material undergoing such a transformation thereafter usually stabilizes, thereby attaining a new property profile. Therefore the potential of polymorphic change has to be investigated when the substance is determined to be processed for dosage forms containing

Table 2: Physical Stability and Methods of Investigation

Physical Stability
Amorphism Polymorphism Pseudopolymorphism (Solvates) Hydrates: Formation Properties
Methods of Investigation
X-Ray diffraction IR-Spectrum in Nujol Thermoanalytical methods Thermomicroscopy DSC TGA Gravimetry Water determination methods Solubility determination Analytical/Equilibrium Synthetic/Quasi-equilibrium

Table 3: Properties of Pharmaceutical Relevance changing with (Pseudo-)Polymorphic State

- Crystal shape
  (Particle Size)
- Density
  Mass (hydrates)
- Thermal Behaviour
  Melting temperature
- Colour
  Refractive Index
- Solubility
  Dissolution Rate
- Mechanical Properties
  Milling
  Compression

the substance as solid particles. The search for possible or existing polymorphs may be conducted in a "passive" manner by monitoring the sequence of several laboratory batch samples provided by the chemist for differences in solid state IR spectra, X-ray powder diffraction diagrams and thermoanalytical behaviour. However, it is better to embark on an active search by recrystallizing samples from a series of standard solvents under a variety of conditions provided that the limitations mentioned previously allow such procedures.

Several properties of pharmaceutical importance depend on the polymorphic state and consequently do change or may change in case of polymorphic transformations (Table 3).

Certain processing steps are especially prone to induce transformations. Exposure to humid air, and more so to liquid water, provokes hydrate formation. This same condition also may trigger stabilizing transformation of amorphous into crystalline forms. Milling and compression can cause polymorphic changes.

It is a good rule to provide a polymorphic active substance as one single and as its most stable polymorph. There may be exceptions to this rule:

1. The thermodynamically most stable polymorph is unsufficiently bioavailable due to unfavourable solubility and dissolution properties.
   Chloramphenicol palmitate exemplifies how large a difference in bioavailability can exist between a stable polymorph (A) and a metastable polymorph (B) [4; 5].
2. Two polymorphic forms may incidentially have the same free energy within the temperature range of pharmaceutical significance and therefore be equal

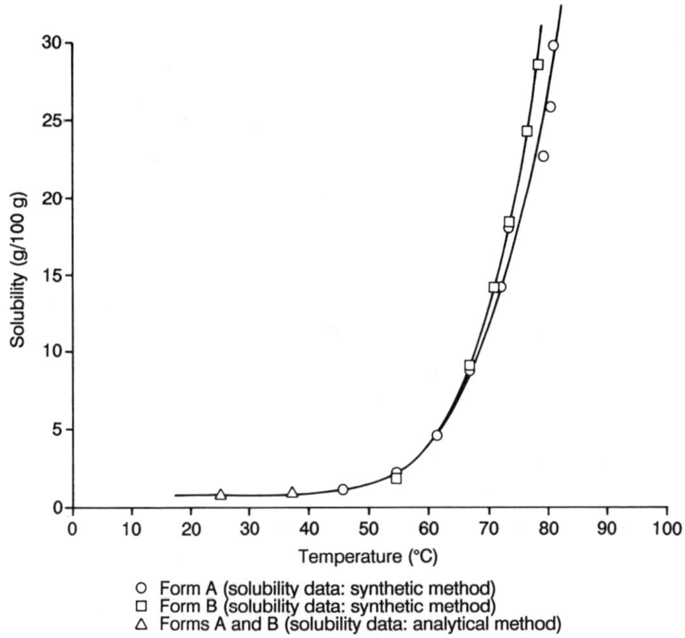

Fig. 1: Maprotiline Hydrochloride: Solubility in water vs. temperature.

also in their biopharmaceutically relevant properties such as solubility and dissolution rate.

An example for the latter case is demonstrated by the temperature curves of two polymorphs of maprotiline hydrochloride (Figure 1). Within experimental errors their solubility at room temperature and up to approx. 70 °C is identical. Only above 80 °C distinct polymorph-specific differences become apparent.

Polymorphic instability can cause serious and hard-to-solve problems. Some time ago we tried to micronize a sparingly soluble calcium antagonist, CGP 28727, in a jet mill, without success, however. Milling in a pin mill reduced the particle size only marginally but the output of the mill was yellow in contrast to the cream coloured material fed into the mill. Chemical decomposition could be ruled out. Surprisingly enough, the melting range of the yellow substance was found to be 170.5–172.5 °C, whereas the cream coloured starting material solidified after initial melting at 155–157 °C in order to melt finally at 172 °C. Thus it appeared that milling had triggered the stabilization of a metastable polymorph (see data Table 4).

As a second surprise, the water solubility of the high melting yellow substance was found to be higher than that of the pale starting material. Moreover,

Table 4: Properties of Two Polymorphs of CGP 28727

Polymorphic Form	**I**	**II**
Colour	cream coloured	yellow
Melting behaviour	First melting at 155–157.5 °C, solidifies, final melting at 172 °C	Melting range 170.5–172.5 °C
Solubility [mg/100 ml solution] water, @ 25 °C	0.38	0.74
Physical stability in aqueous suspension	stable	not stable; transforms into Polymorph I
Milling	not possible; transforms into Polymorph II	possible

the yellow substance turned pale on extended stirring in water. Also upon boiling in water the yellow colour faded rapidly. IR spectra and X-ray diffraction pattern revealed that it was the low melting starting material which was recovered from the aqueous suspensions. Whereas commonly a higher melting point is associated with higher thermodynamic stability and lower solubility, CGP 28727 is an example of the less frequently encountered enantiomorphic case when inversion of the relative thermodynamic stability takes place within the range of pharmaceutical working conditions; the inversion temperature is between 100 °C and 155 °C (which is the melting temperature of the cream coloured form) [6].

2.2 Hydrates

According to our experience, we expect one third of all NCEs to be liable to hydrate formation. This adds further complications to the drawbacks of polymorphism. Water of crystallization *taken up* by the solid a. s. increases its volume and changes the mechanical properties. Water of crystallization *released* from a hydrate may foster degradation as a solvent or as a reactant for hydrolysis. Classically, small samples of a. s. are exposed to stepwise increasing humidity until constant weight is reached; thereafter the sample steps down the stair of humidities to conclude a complete adsorption/desorption cycle resulting in the sorption isotherm. Figure 2 shows the water vapour sorption isotherm of diclofenac sodium as an example. Confining the investigation to water vapour, however, would miss those hydrates which form for the first time in liquid water only. We encountered substances which even in water remained anhydrous for 24 and up to 48 hours but after one week stirring in water had transformed into a hydrate.

Carbamazepine may illustrate complications resulting from hydrate forma-

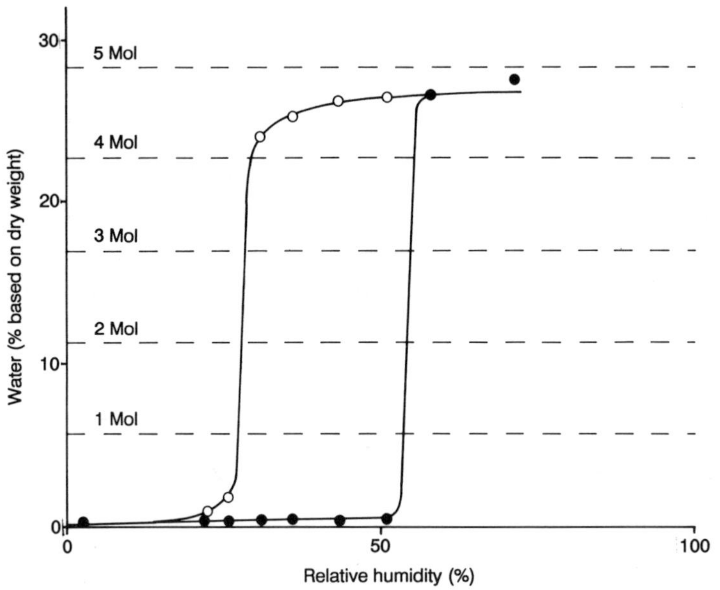

Fig. 2: Diclofenac Sodium: Water vapour sorption isotherm, 25 °C.
● Adsorption, ○ Desorption

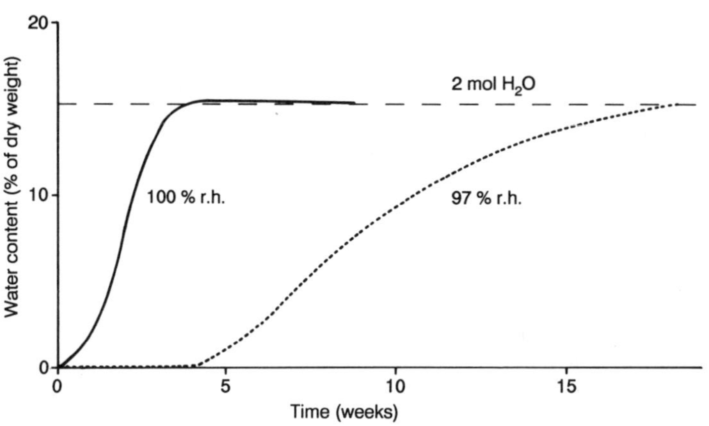

Fig. 3: Carbamazepine: Formation of the dihydrate from anhydrous carbamazepine in humid air at 23 °C. From [7]

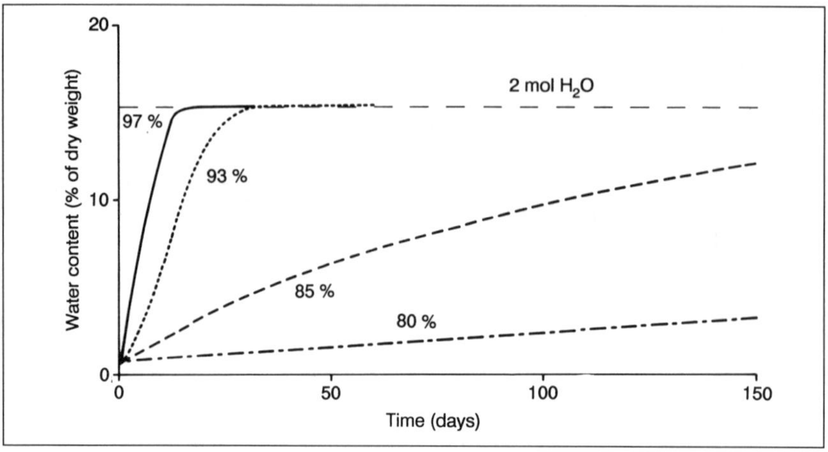

Fig. 4: Carbamazepine: Rehydration of carbamazepine at 23 °C and various humidities after drying the dihydrate to a residual water content of 0.05 mol. From [7]

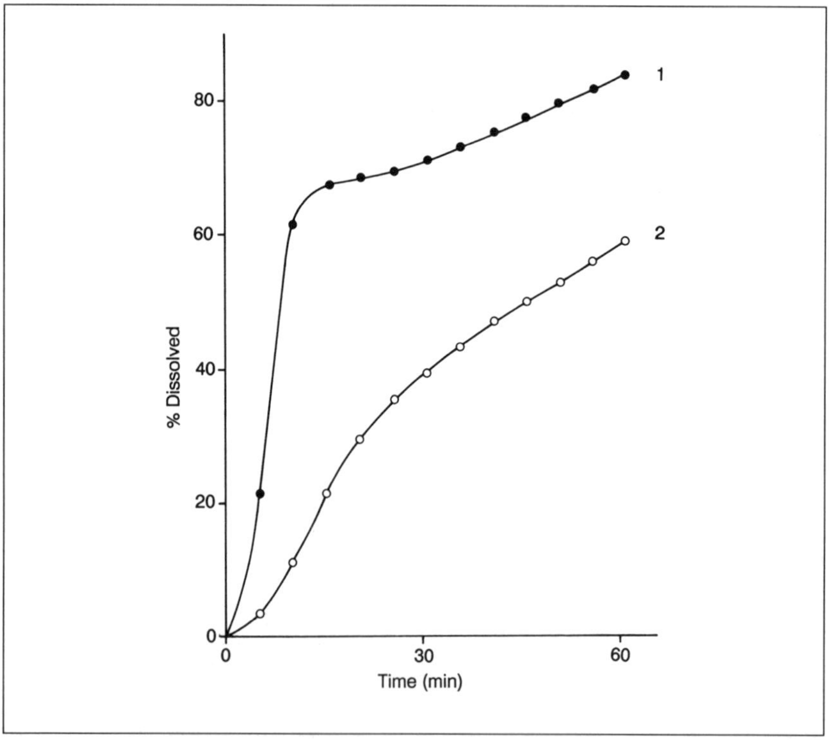

Fig. 5: Dissolution Rate of 500 mg Ampicillin
1 anhydrous ampicillin, 2 ampicillin trihydrate. From [8]

tion. In very humid air, this anticonvulsive slowly forms a dihydrate (Figure 3). In contrast to the slow hydration with atmospheric water vapour it is a matter of a few minutes for the dihydrate needles to start growing from anhydrous carbamazepine in contact with liquid water. As a consequence, lumps of carbamazepine solidify to plaster-like masses when wetted with water or an aqueous liquid. This may happen even to tablets if the formulator has not provided for the disintegration of the tablet to be faster than the hydration process.

If, after wet granulation and (insufficient) drying, traces of the dihydrate are retained then such traces act as seed material, and rehydration takes place faster and would also happen at lower humidities, as shown in Figure 4 [7].

Solubility and dissolution rates of hydrates differ from those of the respective anhydrous species for which the dissolution rate curves of ampicillin in Figure 5 may serve as an example.

Before a hydrate of an active substance is being considered as the final form for manufacturing and processing of dosage forms its physical stability to climatic conditions has to be tested. Although thermogravimetric analysis appears to be the ideal technique and at first glance might give the correct answer by a temperature scan within minutes, it has to be born in mind that the long term stability can be verified by the isothermal situation only. All climatic and thermal conditions occurring during manufacture, packaging and storage should be anticipated by a hydrate challenge program. Table 5 was designed to assist in the decision whether a hydrate or the anhydrous form of a given active substance should be chosen for the development of solid dosage forms where such questions are particularly crucial.

Particle size reduction may cause stability problems. Three frequently encountered problem areas are: (1) polymorphic changes during milling; (2) caking and formation of smaller agglomerates or large lumps during storage may annihilate the milling results; (3) activation and increased surface area may accelerate solid state degradation.

2.3 Chemical Stability

There are several ways by which an instable drug substance can undergo decomposition (Table 6). Within the preformulation program, the stability of the substance as a solid and in solution is tested according to the schedules given in Table 7–9. The stability of the solid substance is tested within the compatibility program. As a rule, the preferred method for sample analysis is qualitative TLC. For critical cases, densitometric quantification is applied. The use of HPLC is rather an exception at this stage of development. Visual changes of the stressed samples are also recorded.

Substances with critical stability require more attention with respect to the conditions causing degradation. In particular, the pH and temperature dependent degradation kinetics are studied quantitatively in order to decide on such

Table 5: Formation and Stability of Hydrates of Active Substances Consequences for Solid Dosage Forms

Formation of Hydrate		**Stability of Hydrate**				**Use Form of Substance** A = Anyhydrous H = Hydrate		
Hydrate is formed in water and …		R.H. [%]	Temperature [°C] 25	35	50	Manufacture and Storage	Present in Solid Dosage Form after Processing: dry	wet
… at low humidity		11		Hydrate stable		H	H	H
… at intermediate humidity		22				A	A	H
… at high humidity		33				A	A	A
… in water only		43		Hydrate not stable		A	A	A
No hydrate formed						A		

Hatched: Protect from moisture and humidity! Dotted: Protect from moisture and humidity! Unfavourable form of substances!

Table 6: Chemical Stability

Degradation Reactions:	• Hydrolysis • Oxidation • Radical Reactions • Rearrangement Reactions • Isomerization • Racemization • Epimerization • Dimerization, Polymerization • Decarboxylation • Reaction with an inactive moiety (in case of salts, complexes etc.)

questions as, e.g., the feasibility of injectable solutions or lyophilisates, and degradation during gastrointestinal passage.

Light stability is evaluated in view of the usual light exposure during handling, manufacturing and processing including analytical work. For exposure to a high energy light source (Xenotest; light intensity: 200000 lux) special flat-bed powder cuvettes with a defined bed depth of 0.25 mm have been designed [6]. Thus, specifically high and reproducible changes by light irradiation as well as easy visual assessment of discolouration is achieved.

Stability considerations also include interactions of the active substance with excipients and other materials (Table 10). Drug/excipient compatibility testing is addressed in a separate contribution. Therefore it is not dealt with here but just mentioned briefly. We had experienced pitfalls a few years ago when we still relied exclusively on the testing of binary drug/excipient mixtures. Thereupon we decided to include two standard mixtures in the compatibility program. One of them represents a typical wet granulation type formula, the other one a dry compression type formula (Table 9). But still the binary drug/excipient mixtures are regarded to be a valuable first quick screen to rule out grossly incompatible excipients.

Another important interaction refers to processing tools. Weak drug bases are commonly turned into water soluble salts by using strong acids. Occasionally the acidity of the strong acid is only insufficiently balanced by the weak base. A low pH of the saturated aqueous solution of an acidic salt and hygroscopicity may combine to corrode the polished steel surfaces of tabletting tools. On the other hand, by such a constellation metal traces may get onto the surface of tablets compressed from material containing corroding salts. This causes, for example, catalyzed degradation or discoloured specks. According to our experience it is advisable to test the corrosiveness of salts whenever the pH of the saturated aqueous solution is 3 or less. The substance is placed on mirror-polished sample disks machined from the essential types of tooling steel and exposed to a controlled atmosphere of 75% r.h. for 1 h. Removing the substance by dry sweeping, then rinsing with water and drying reveals any kind of

Table 7: Determination of Properties of the Active Substance in Solution

	Electrolyte		Nonelectrolyte
Electrolytic Properties	Ionizable Groups Type and number: acid; base; salt ampholyte zwitterion Strength: pK_a-values		—
Solubility	Aqueous solvents water 0.9% aqu. NaCl-solution simulated gastric fluid simulated intestinal fluid		
	Profile pH = 1...10 (aqueous buffers)		—
	Nonaqueous solvents ethanol methanol dichloromethane PEG 300		
Partition	n-octanol / aqu. phosphate buffer pH = 7.4		
	Profile pH = 1...10 (aqueous buffers)		—
Stability	Solvent		Nature of stress
	Aqueous solvents water aqueous buffers	pH and	heat oxygen light
	Selected nonaqueous solvents		

corrosion [6]. The data of two corrosive salts (Table 11) show that the key parameter for corrosiveness is the pH of the saturated aqueous solution and not the pK_a of the parent drug base.

In case of positive findings, it depends on the severity of corroding attack and on the intended dosages whether the salt still can be considered as a useful form of the active substance. If the dosage allows appropriate dilution to "neutralize" the corrosiveness this negative property can be managed, or else a different chemical form of the substance must be found.

Table 8: Stability of the Active Substance in Solution

Concentration	(0.05 …) 1 g/100 ml		
Solvent	Aqueous (up to 50 % by vol. EtOH if necessary)		
	pH	Flushing Gas	Stress (details below)
water	unbuffered	O_2; N_2	Heat, Light
0.1 N HCl 0.1 AcOH buffers: acetate phosphate borate (0.1 N NaOH)	1 3; 4.5 6; 7.4 12	none (= air)	Heat

Stress	**Temperature** [°C]	**Exposure time**
Heat	25, 50, 80	2 d, 1 w
	100	2 h
	120	20 min
Light	approx. 30	6 h, 24 h

Table 9: Stability in the Presence of Excipients: Testing Chemical Compatibility

Step	**Samples**	**Stress Conditions *)**			
		Temp. [°C]	1 w	3 w	9 w
①	Active Substance (AS) Binary mixtures with excipients (EX) AS+EX = 1+1 … 1+999 Multiple mixture A Multiple mixture B	25			
		50			
		65	+		
		80	+		
	Active Substance (AS) Multiple mixture A	approx. 30	Light (Xenotest) 6 h, 24 h		

*) Samples of dry mixtures each in sealed vials and in open vials at 75% r.h. and temperatures as specified

Table 9: Stability in the Presence of Excipients: Testing Chemical Compatibility

Step	Samples	Stress Conditions *)			
		Temp. [°C]	1 w	3 w	9 w
②	Active Substance Additional multiple mixtures (C, D, E ...)	25			+
		35			+
		50		+	+
③	Prototype Formulations	65	+	+	
		80	+		

*) Samples of dry mixtures each in sealed vials and in open vials at 75% r.h. and temperatures as specified

Table 10: Interactions of Drug Substances with Other Materials

Excipients	
physical	Solubilization Complexation Absorption Adsorption
chemical	Reaction Catalysis Inhibition
Container components **Processing equipment materials**	

physical	Sorption
	  Adsorption
	  Absorption
	Migration

chemical	Corrosion

Table 11: Properties of Corrosive Active Substances

Substance	**CGP 21495 A**	**CGP 8173 A**
Type of salt pK_a of drug base	Hydrochloride 2.92 ± 0.05	Methane sulfonate 6.2 ± 0.1
Solubility in water at room temperature	1.24 g/100 ml solution	78 g/100 g solution
pH of the saturated aqueous solution	2.1	2.2

3. Stabilization

As great as the variety of lurking stability problems is, as widely vary the means for stabilization and their points of action. Table 12 gives an overview on the general principles.

The selection of the optimum chemical form is a crucial step oftentimes also in the sense of a preventive means of stabilization. When the definitive salt form of an antidepressive compound, CGP 15210, had to be decided upon from a set of three available salts (A, B and C; Table 13) the fumarate (B) appeared to have a good profile of properties, and it was proposed as the top candidate. However, after recrystallization of some 100 grams of the fumarate salt the TLC revealed that an impurity previously present in traces only had increased to a level of several per cent. It was identified as the product of covalent addition of the secondary amine to fumaric acid. This finding of course gave reason to discard the fumarate and to try further acids for salt formation.

Table 12: Stabilization of Drug Substances: Bulk Material

Stabilization by	
Exclusion of destabilizing principes	
Chemical	Selection of chemical form Solvents Heavy metal ions Other impurities
Environmetal	Heat Moisture, humidity Oxygen Carbon dioxide
Addition of stabilizing principles	
	Complex forming agents Antioxidants Crystallization modifiers
Process Management	
Modulation of existing process	Avoiding conditions with destabilizing effect Reduction of specific surface area increasing particle size Reduction of amorphous proportion/ avoiding amorphism Increasing purity
Stabilizing processing steps	Additional processing steps e.g. coating, microencapsulation, complexation etc.

Table 13: CGP 15210: Properties of Base and Various Salts

Substance	CGP 15210 Base	CGP 15210 A Fumarate	CGP 15210 B Oxalate	CGP 15210 C Mesylate	CGP 15210 D Succinate	CGP 15210 F Hydrochloride	CGP 15210 G Sulfate
Molecular Formula	$C_{13}H_{19}NO_2$ (= Base)	$(Base)_2 \cdot C_4H_4O_4$	$(Base)_2 \cdot C_2H_2O_4$	$Base \cdot CH_4O_3S$	$(Base)_2 \cdot C_4H_6O_4$	$C_{13}H_{19}NO_2 \cdot HCl$	$(Base)_2 \cdot H_2SO_4$
Rel. Molar Mass	221.30	558.67	532.64	317.40	560.69	257.76	540.68
Melting Range, °C	117–118	190–192	245–250	138–140	150–152	214–215	247; 270–271
Solubility in	g/100 ml	g/100 ml solution	g/100 ml solution	g/100 g solution	g/100 g solution	g/100 g solution	g/100 ml solution
water, 25°C (pH)	<0.1	2.63 (5.4)	~0.5	54 (3.7)	57–58 (6.3)	56–57.5 (1.9)	2.15 (5.3)
simulated fluids							
– gastric, 37°C (pH)	–	–	–	61–62 (0.6)	60–61 (6.1)	68–70.5 (0.2)	4.63 (1.1)
– intest., 37°C (pH)	–	–	–	60–63 (5.2)	43–48.5 (6.3)	68–71 (5.1)	3.23 (6.6)
0.1N HCl, 25°C (pH)	–	5.44 (3.7)	–	–	–	–	–
phosphate buffer pH=7.4, 25°C (pH)	–	3.30 (7.1)	–	–	–	–	–
Hygroscopic Behaviour (23°C)	–	Little uptake of water up to 92% r.h. (1.4–2% water); probably formation of a hydrate at 97% r.h.; hydrate water lost at $\varphi \leq 80\%$ r.h.	–	No change at rel. humidity ≤ 54%. Formation of a monohydrate above 54% r.h. which looses water only at 50°C/12% r.h.	No change up to ~40% r.h.; hygroscopic at >43% r.h.; above 75–80% r.h. possibly hydrate formation; deliquesces at >97% r.h.	Water uptake between 43 and 75% r.h.: 0.3%; between 80 and 93% r.h.: up to 2 mol water; deliquesces above 93% r.h.	Forms a dihydrate
pH of a 1% aqueous solution	–	–	–	5.5	6.6	5.4	–
Additional remarks	–	Not stable in two different tablet model formulations	–	–	–	–	Stable in two different tablet model formulations

Finally the neutral sulfate salt (G) satisfied all requirements including good stability.

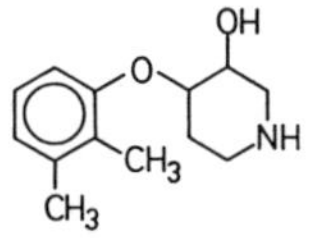

CGP 15210

Identification of the most suitable chemical form for a NCE can reach far back into the realm of the pharmaceutical research chemist, and it should be the goal of the development pharmacist to be involved with equal voice in the team of chemists and biologists when the final candidate is going to be selected from a group of biologically equivalent alternatives or, for example, from a series of prodrugs, thereby duely taking into account stability considerations (Table 14).

Additives may be tried to stabilize a labile bulk drug. Of course, such additives can only be taken from the series of pharmaceutically accepted materials and within reasonable limits. As judged from the result of literature and patent searches, it appears that stabilization of labile compounds by additives either is merely a theoretical matter, or else and more likely it is preferably kept confidential as know-how on the manufacturer's side. In any case, stabilization by additives is typically a matter of the developing pharmacist's, and for the sake of purity of the bulk drug the chemist should resort to additives as a "last aid" only.

Apart from crystal polymorphism, habit and size of crystals can also contribute to or alleviate stability problems mainly with drugs in suspensions. Accordingly, the crystallization step may be critical. Many parameters of the final crystallization or precipitation process must be fine tuned for the desired result. Here the possibility should be mentioned to take influence on the process, again by traces of additives, for shaping the resulting crystals without altering the polymorphic state [9]. It is well known that from water, rock salt crystallizes as cubes. But with urea added, sodium chloride octahedra grow from the solution. By applying such principles, carbamazepine dihydrate which normally grows as needles measuring up to centimeters in lengths can be "domesticated" to form cubes. Since during morphological crystal engineering by

Table 14: Stability considerations: Selection of Chemical Form of Active Substance

• Acid/Base	
• Salt	
• Resinate	
• Complex	
• Derivative	Acid, Ester, Amide
	Alcohol, Ester
	Acid, Alcohol
	Prodrug, Metabolite

Table 15: Active Substance Stability I

Time \ Temperature	−15 °C	8 °C	25 °C	30 °C	40 °C	50 °C	25 °C 75% r.h.
Initial			G C				
3 weeks							G C
3 months	C	C	G C	C [G]	G [C]	G	
6 months	C	C	G C	C [G]	G [C]	G	
1 year	C	C	G C	C [G]	G [C]		
2 years	[C]	[C]	[C G]	[C G]			
3 years							
4 years							
5 years							

Active Substance Stability II: First Batch

Time \ Temperature	−15 °C	8 °C	25 °C	30 °C	40 °C	50 °C	25 °C 75% r.h.
Initial			G C				
3 weeks							G C
3 months	C	C	G C	C [G]	G [C]	G	
6 months	C		C	G C	G [C]	G	
1 year	C	C	G C	C [G]	G [C]		
2 years		C	G C	[G]			
3 years	C	C	G C	[G]			
4 years		C	G C				
5 years	C	C	G C				

Active Substance Stability II: Second and Third Batch

Time \ Temperature	−15 °C	8 °C	25 °C	30 °C	40 °C	50 °C	25 °C 75% r.h.
Initial			G C				
3 weeks							
3 months		C	G C				
6 months		C	G C				
1 year		C	G C	C [G]			
2 years		C	G C				
3 years		C	G C				
4 years			G C				
5 years	C	C	G C				

Legend:
- G Products with good stability expected
- C Products with critical stability expected
- [ ] Additional facultative storage conditions / sampling times

adsorption-inhibition additives, minute traces of the face-occluding modifiers are included into the crystals they must equally comply with compendial and regulatory requirements.

4. Stability characterization of the active substance

Once the chemical form of the active substance and the process for its manufacture have been established and the first batch has been prepared, analyzed and released, then stability tests on the active substance are initiated by Chemical Development. The stress and sampling schedule for an active substance is shown in Table 15. "Stability I" is a pilot type stability schedule serving to identify the normal storage requirements of the bulk drug; "Stability II" is designed to document the stability of batches used for clinical trials up to the maximum expiry date. The increasing stability experience with a given substance enables to stepwise reduce the work load of stability testing. The resulting data serve to satisfy the regulatory requirements.

5. Stabilization of the active substance in dosage forms

The pharmacist's mission in development is to design drug application forms and procedures for their production. An essential element of this activity is the goal to keep all chemical and physical parameters of the dosage form and of its

Table 16: Physical Stabilization of Active Substances in Dosage Forms (1)

State of A.S.	Inhibit change	by
Solid	Modification and habit ("Tracht")	Preference of stable polymorph Choice of excipients, additives
	Evaporation, sublimation	Complexation Microencapsulation Coating
	Hygroscopic deliquescence	Ab-/adsorbing excipients Complexation Microencapsulation Nonaqueous coating
Solid, dispersed	Particle/crystal growth	Choice of dispersant Avoiding – (super-)saturation – (excessive) temperature changes
Dissolved	Precipitation, crystal growth	Avoiding – (super-)saturation – nucleation
Adapted from W. Erni (1982)		

Table 17: Physical Stabilization of Active Substances in Dosage Forms (2)

State of A.S.	Avoid Recrystallization/ Transformation by inhibition of	Measure
Metastable	– nucleation	appropriate conditions of – processing – storage
	– growth of nuclei	decreasing the proportion of liquid phase reduction of moisture dilution with excipients
		increasing viscosity addition of: structure builders or structure breakers tensides crystallographically related substances
Adapted from C. Führer [10]		

performance invariable along the time axis. In particular, the integrity of the drug substance has to be maintained.

The stabilization principles employed during dosage form development can be dealt with here in very general terms only. Again, the stabilization refers to either physical parameters or to the chemical integrity of the active substance. Of course, both areas are interrelated to some extent. In Tables 16–18 the various stabilization principles applicable in dosage form development are compiled. Special attention deserve the cyclodextrins and their highly water soluble derivatives. As hosts capable of forming molecular inclusion complexes with such guests as drug molecules of appropriate dimensions and properties, they are not only potent solubilizers but also in many cases have been shown to effect chemical and physical stabilization of their guest molecules as well. This principle of stabilization is dealt with in a separate contribution in this volume.

Table 18: Chemical and Microbial Stabilization of Active Substances in Dosage Forms

Chemical Stabilization				
Measure	**Hydrolysis**	**Oxidation**	**Light**	**Heat**
pH-Adjustment	+	+		
Moisture exclusion/adjustment	+			
Cosolvents	+			
Microencapsulation/Coating	+	+		
Micellar inclusion	+	+		
Molecular inclusion	+	+		
Complexation of A.S.	+	+		
Complexation of metal traces		+		
Displacement of oxygen (N_2, Ar)		+		
Antioxidants and synergists		+		
Opacity			+	
Packaging			+	+
Storage				+
Microbial Stabilization				
Stabilization against	Microbial degradation			
Prevention of	Microbial infection			
Adapted from W. Erni [11]				

References

[1] (a) T. Fujita, J. Iwasa, C. Hansch, J. Amer. Chem. Soc. 86 (1964), 5175–5180; (b) G.C. Nys, R.F. Rekker, Chim. Ther. 8 (1973), 521; (c) C. Hansch, A. Leo, Substituent Constants for Correlation Analysis in Chemistry and Biology. John Wiley & Sons, New York etc., 1979

[2] (a) D. Mackay, W.Y. Shiu, J. Chem. Eng. Data 22 (1977), 399–402; (b) S.C. Valvani, S.H. Yalkowsky in: Physical Chemical Properties of Drugs, S.H. Yalkowsky, A.A. Sinkula, S.C. Valvani (editors), Marcel Dekker Inc., New York, Basel, 1980, pp. 201–229; (c) S.H. Yalkowsky in: Techniques of Solubilization of Drugs, S.H. Yalkowsky (editor), Marcel Dekker Inc., New York, Basel, 1981, pp. 1–14

[3] D.D. Perrin, B. Dempsey, E.P. Serjeant, pK_a Prediction for Organic Acids and Bases, Chapman and Hall, London, New York, 1981

[4] A.J. Aguiar, J. Krc, A.W. Kinkel, J.C. Samyn, J. Pharm. Sci. 56 (1967), 847–853

[5] A.J. Aguiar, J.E. Zelmer, ibid. 58 (1969), 983–987

[6] P.H. Stahl, in: Stabilisierungstechnologie. Wege zur haltbaren Arzneiform, D. Essig, J. Hofer, P.C. Schmidt, H. Stumpf (editors), APV Paperback Vol. 15, Wissenschaftliche Verlagsgesellschaft Stuttgart 1986, pp. 15–32

[7] P.H. Stahl in: Towards Better Safety of Drugs and Pharmaceutical Products, D.D. Breimer (editor), Elsevier/North Holland Biomedical Press, Amsterdam etc., 1980, p. 265

[8] J.W. Poole, G. Owen, G. Silverio, J.W. Freyhof, S.B. Rosenman, Curr. Therapeut. Res. 10 (1968), 292–303

[9] L. Addadi, Z. Berkovitch-Yellin, I. Weissbuch, J. van Mil, L.J.W. Shimon, M. Lahav, L. Leiserowitz, Angew. Chem. 97 (1985), 476–496

[10] C. Führer, in: Stabilisierungstechnologie. Wege zur haltbaren Arzneiform, D. Essig, J. Hofer, P.C. Schmidt, H. Stumpf (editors), APV Paperback Vol. 15, Wissenschaftliche Verlagsgesellschaft Stuttgart 1986, pp. 33–43

[11] W. Erni, GSIA-Fortbildungskurs 1981/82, Basel, 28JAN82

3 Excipient compatibility possibilities and limitations in stability prediction

Donald C. Monkhouse, USA-King of Prussia

The study of solid state reactions is fundamental to the development of a new drug candidate destined to be marketed as a solid dosage form. Kinetic studies must be undertaken to gain a basic understanding of the reaction mechanism so that stability of the drug can be predicted. From such information, overage determinations, label statements, and expiration dating are ultimately evolved. Once formulated, however, a drug molecule rarely exhibits the same mechanism of degradation as in its pure state. The addition of excipients only further exaggerates heterogeneity already common to the drug substance per se. Many factors such as moisture, pH of the excipient surface, and volume of the container can further affect the stability behavior of the active ingredient. Free or bound granulating solvents or water, differences in thermal capacity between drug and excipients, phase changes, and chemical equilibria also contribute to the complex nature of formulation stability studies. The question to be addressed in this paper is the justification of conducting stability studies on complex drug-excipient mixtures in order to select the most stable functional grouping, salt form, polymorph or formulation composition.

The industrial research pharmacist must primarily concern him/herself with the initial degradation of a formulation. Rapid decomposition of a formulation under reasonable conditions of storage is clearly unacceptable for a commercial product. Preformulation kinetics in these cases take on a more practical value: many solid state reactions, even those of a formulated drug, are amenable to Arrhenius testing during the first 5–10% decomposition, even though an examination of their complete degradation profile might reveal a complex exponential or S-shaped curve. Careful interpretation of such Arrhenius data can provide adequate prediction of stability without demanding the complete elucidation of a reaction mechanism. Furthermore, it is the contention of the author that such studies should be performed on the complete formulation rather than being limited to single drug-excipient combinations. Only if a major incompatibility is suspected should individual drug-excipient mixtures be examined in order to remove the offending ingredient.

Over the past 30–40 years the practice of setting up prospective complex drug-excipient mixtures as accurate predictors of instability has been accepted without question. Only in the best of cases could a binary mixture serve as a crude representation of what might occur in a multi-ingredient formulation. In modern drug development where time is of the essence, there is a great economic incentive to formulate a stable dosage form early in the process. The purpose of this discussion is to gain concurrence from the reader that, in many cases, incompatibility of excipients can be predicted from the literature, thus saving the formulator and the corporation much valuable time in the development process. Alternative strategies to obtaining a stable formulation will be offered.

Stability studies are performed to assure safety, efficacy, purity and quality of a new drug product. From a safety perspective it is important to minimize chemical instability in order that humans are not exposed to new degradation products that were not present for the original toxicology studies. Also, any physical instability can adversely affect release rate thereby affecting clinical performance. Efficacy is clearly important, to ensure, through potency retention and impurity profiles, that the product must meet specifications for levels of degradation throughout its lifetime. The overriding consideration is that quality does not change over time, so that the product remains suitable for human consumption. A recent survey indicated that instability was by far the leading reason for termination of clinical trials during a 15 year period, thus illustrating the commercial value of such studies.

There are basically two categories of stability studies – exploratory and confirmatory. Exploratory studies are conducted to support physicochemical characterization and aid in the preliminary elucidation of degradation mechanisms. These early studies establish kinetics of degradation, the various factors affecting degradation (such as moisture, oxygen, light and metal ions), degradation product identification, stability indicating assay methodology (mass balance) and particle size specifications (for homogeneity, bioavailability assurance and stability). Based on the information gained, prototype formulations are developed using excipients that are predicted to be compatible with the drug. Initial stability studies are then conducted on the preliminary dosage form. If any instability is noticed, it is not too late at this stage to conduct drug-excipient compatibility studies to identify incompatible excipients (or mixtures thereof) in order to eliminate them.

Confirmatory studies, on the other hand, are formal protocol-driven studies that support regulatory filings. Results from these studies are used to establish extent of degradation to be expected during the lifetime of the product, shelf-life prediction and expiration dating. The goal for such studies is to avoid, at all costs, creating data which is unreportable and which cannot be explained. Statistics show that such data can inexplicably hold up regulatory approval of new drug products. Clearly then, rational stability protocol design requires an understanding of the physical and chemical properties of the drug and its accompanying excipients. Any surprise found in a confirmatory stability study should be inexcusable if the proper exploratory studies are conducted with care and intelligence.

Most formulation science practitioners in the pharmaceutical industry would agree that traditional compatibility studies have shown limited value in streamlining product development. Scientific deficiencies of such studies include the fact that too few variables are addressed, unrealistic excipient-to-drug ratios are studied leading to misleading or ambiguous results and processing effects are often ignored. From a business perspective they are uneconomical since the time lost could be better allocated to studying prototypes rather than binary mixtures. This inefficiency is further compounded if the results are inconclusive.

A suitable alternative would be the "rational" approach to preformulation studies through consideration of the structure of the drug, its site of delivery, the desired characteristics of the dosage form and a working knowledge of the chemistry of the drug substance and reactivities of available excipients. In this way, undesirable reactions resulting from inappropriate selection of excipients can be prospectively ruled out.

A good working knowledge of the reactivity of excipients can be found in the literature [1]. In 1954, Blythe and Zirkle were quoted as saying, "Pharmaceutical manufacturers have become increasingly aware of the need for early introduction of their products to the physician. One development problem is the stability of the product, and one way of speeding the solution of this problem is use of the literature".

In reviewing the history of compatibility testing, it was recognized early on that pharmaceutical stability was adversely affected by the presence of excipients. Earliest tests focussed on examining the physical appearance of binary mixtures to look for a color change. As analytical techniques improved, the way of reporting changes became formalized and more sophisticated (quantitative rather than qualitative). Over the years, drug excipient compatibility testing became dogma for the industry with no real evidence that the procedures employed actually accomplished the task. Neither is it too surprising that there is a dearth of research articles describing instances where compatibility testing was ineffective.

In the last decade, manufacturing processes and formulations have become increasingly more complex but the old ways of screening drug excipient incompatibilities have not kept up with the sophistication of science in the industry. The classic destabilizing excipients – magnesium stearate, stearic acid and sodium lauryl sulfate – have not changed much, although their purity has improved to the extent that the impurities (e.g., metal ions) which may have served as catalysts for degradation have been substantially reduced. Furthermore, the panacea of finding an excipient that actually stabilizes a dosage form has yet to be realized (although polymers, antioxidants and cyclodextrins have been proposed).

Expanding further on this last point, it is a myth to expect an additive to stabilize a solid state reaction. This is because in contrast to solution state reactions which are homogeneous and occur from molecular interactions which are spontaneous, solid state reactions are generally heterogeneous reactions which occur only at points of contact between excipients and drug substance.

Known Reactivities:

Functional Group	Incompatibilities	Type of Reaction
primary amine	mono and disaccarides	amine-aldehyde and amine-acetal
ester, cyclic, lactone	basic components	ring opening ester-base hydrolysis
carbonyl, hydroxyl	silanol	hydrogen bonding
aldehyde	amine carbohydrates	aldehyde-amine Schiff base or glycosylamine formation
carboxyl	bases	salt formation
alcohol	oxygen	oxidation to aldehydes and ketones
sulfhydryl	oxygen	dimerization
phenol	metals	complexation
gelatin capsule shell	cationic surfactants	denaturation

To expect a stabilizer to interpose at these points of contact on a random basis is rather simplistic.

Other pitfalls include the fact that the degree of crystallinity of the drug substance can markedly affect stability just as it does with the presence of free or bound moisture (or granulating solvent) which affects the pH of the surface.

Rational design, rather than empirical testing, is based on correctly predicting incompatibilities and the judicious selection of excipient blends. Current training in the pharmaceutical sciences teaches that some known reactivities are as shown in the following table:

Sources of information on excipient incompatibilities may include scientific journals, bulletins, trade magazines, patents, government reports, theses, texts, reference books, manufacturers technical specifications and personal communications with consultants. One particularly useful reference is the *Handbook of Pharmaceutical Excipients* which contains a section on "incompatibilities" in each monograph [2]. Particularly noteworthy examples include those of lactose (browning reaction), magnesium stearate (acid and bases) and sodium lauryl sulfate (cationic substances).

There are four major attributes of a drug-excipient compatibility study: 1. sample preparation, 2. statistical design, 3. storage conditions and 4. method of analysis.

For sample preparation, the drug and excipient can be slurried in a solvent for the drug e.g., isopropanol, chloroform, acetone, methylene chloride or even water, or mixed dry in various ratios via trituration, ball milling or turbular mixer. Excessive grinding can, however, reduce the degree of crystallinity and increase reactivity of the drug substance. Alternatively, the mixture can be compacted, e.g., in a Carver Press, to simulate forces encountered in a tabletting operation. Various authors [3–7] have recommended specific drug: excipient ratios. For lubricants the ratios vary from 5:1 to 50:1, while for most other diluents, the ratio generally varies from 1:1 to 5:1. The ratios are meant to reflect those expected to be present in the final product. Of course, potency of the drug plays a pivotal role in this decision, e.g., for a 750 mg strength tablet, there would be virtually no excipients used; whereas, for a 0.1 mg strength, the dosage form would consist mainly of excipients and very little drug.

Statistical designs for conducting such studies are invariably employed to reduce the number of samples to be tested. Full to partial factorial designs have been used, usually employing selective regression analysis to identify the optimum excipients and their ratios.

Storage conditions vary widely in temperature and humidity. Depending on whether or not the sample is sealed, for example, in an ampoule, the headspace can vary tremendously in moisture vapour content. Such excipients as povidone, starch, and pregelatinized starch contain equilibrium moisture contents at 25 °C of 12.2, 8 and 7.8% respectively [8] which readily vaporize on heating. Moisture vapour is generally much more reactive than liquid adsorbed onto the surface of an excipient. Of course, if the sample is unsealed, there is no vapour pressure and a falsely encouraging result could be obtained.

Methods of analysis can vary considerably, ranging form thermal techniques, spectroscopy, HPLC and microscopy to solid state NMR. The advantages and disadvantages of each method have been discussed in the literature [9]. A recent paper compared and contrasted Differential Scanning Calorimety (DSC) methodology and Fourier Transform IR for studying the incompatibility between aminophylline and lactose mixtures [10]. DSC can also be used to study the preformulation of proteins. By repeat scanning of thermal events during heating and cooling, the folding and unfolding of the native state can be studied as well as the effect of additives on the protein. When such studies are correlated with bioassays, a powerful predictive tool is created for formulation development of these labile, heat sensitive molecules.

Perhaps the most promising technique to emerge in the recent past to study solid state reactions is that of heat conduction calorimetry [11, 12]. What is particularly attractive about this technique is the ability to study thermal events (exo or endo) at room temperature (accelerated temperature studies can often lead to a change in mechanism from that observed at room temperature). Up until now, instrumentation has not been sensitive enough to measure heat changes of just a few calories relating to 1–2% degradation per year. If correlated to degradation, this technique can be used to distinguish stable **vs.**

unstable formulations. Advantages include the following: (1) sample treatment is not required, (2) the method is non-destructive (allowing other confirmatory analytical methodology to be employed post-study) and (3) the rate of the reaction is measured directly. Whether or not the methodology is useful for solid state reactions where a protective layer is formed and the reaction slows down remains to be seen.

Microcalorimetry is particularly useful in studying batch to batch variations of reactive drug substances such as various cephalosporins [11] (the greater the heat of crystallization, the more stable the substance) and lovastatin [12] (which is oxidized autocatalytically). Here, trace impurities, degree of crystallinity and surface area were shown to be controlling factors for stability prediction.

In the study of drug-excipient interactions, the formulation scientist should be acutely aware of the mechanisms of degradation – either active sites or nuclei, free vs. bound moisture effects, imperfections in crystals and the deleterious effects of high shear processing.

Several case histories (either experienced in-house at SB or reported in the literature) are useful in illustrating how drug-excipient incompatibility studies can give confusing or misleading results:

SK & F 86002: Results were shown to be uninterpretable because of poor quality in homogeneity of the drug excipient sample mixtures prepared.

SK & F 39162: High density of the drug resulted in heterogeneity in sampling, rendering the results extremely variable, and therefore useless.

SK & F 86466: This amine hydrochloride had a melting point of 270 °C and was perfectly stable as free drug. When mixed with a basic excipient the salt dissociated resulting in an oily free amine which volatilized on heating.

Unidentified Enzyme: When loosely mixed with excipients, the enzyme appeared perfectly stable. When compressed with abrasive excipients, however, the frictional heat generated was sufficient to degrade the labile substance.

Captopril: Drug-excipient compatibility studies showed decomposition in the presence of magnesium stearate. Subsequent experience demonstrated that this was an erroneous prediction. Other studies showed that the drug was quite stable in the presence of excipients at a 100 mg strength. However, the drug oxidatively decomposed in a 2 mg strength formulation which had a much lower drug-excipient ratio.

Enalapril [13]: Mixing with microcrystalline cellulose caused dissociation of the amine maleate (weak acid/weak base) which resulted in wicking of the oil away from the anion. The free amine was subsequently proven to be unstable as demonstrated by loss of heat of fusion and loss in weight (thermogravimetric analysis). In the presence of calcium phosphate, no such preferential adsorption occurred and the drug was perfectly stable in the presence of this excipient.

Ibuprofen [14]: This drug forms eutectics in the presence of stearates which then sublime. Rather than chemical instability, it was simply a physical weight loss which accounted for a loss in potency. Film coating the tablets subsequently solved this problem.

Moexipril [15]: This n-carboxylalkyl dipeptide undergoes intramolecular aminolysis in solution at pH's less than 4.5 and ester hydrolysis at pH's greater than 10. In the dry state most excipients were shown to be incompatible; moisture and basic agents were determined to be the dominant destabilizing factors. In wet granulations, however, basic agents were found to suppress drug degradation even in the presence of moisture. A possible explanation for such anomolous behaviour is that a new salt form was created through ion exchange during the wet granulation process.

To reiterate, compatibility studies often prove deficient in accurately predicting stability because of the poor quality of the binary mixture models and the fact that processing effects are often not taken into account. Furthermore, they are not cost effective because they consume valuable resources and can cause inefficiencies in development time.

Perhaps a better alternative is to prepare tablets (or capsules) with standardized excipient blends which have been demonstrated (in-house) to perform trouble free under production conditions. Use of a compaction simulator also ensures that the correct representative forces of compression are applied to the dosage form. In this way a minimal amount of drug is consumed and processing effects (mixing, granulation, compression) are representatively taken into account, as are multiple interactions. By reproducibly compressing single tablets, time is saved and sampling errors are minimized. In this way, development can be considerably optimized. Only if the dosage form proves to be unstable should a retrospective diagnosis be performed to isolate the offending excipient. Excipient blends that are lactose and magnesium stearate "free" are invariably successful in such an exercise.

In conclusion, if the goal is timely development of a stable solid dosage form, then the classic study of binary mixtures of drug and excipients can often prove to be a crude predictor at best. Rather, it is recommended that a more efficient approach is to study compacts of excipient blends with the drug at the desired potency level. In this way, more representative formulations (which are chemically stable and more likely to be viable for the duration of clinical trials) can be chosen for further development. Although this approach contravenes conventional wisdom, it is a small gamble worth taking in a highly competitive drug development environment.

References

[1] R.H. Blythe and M.S. Zirkle, "A Key to Pharmaceutical and Medicinal Chemistry Literature", American Chemical Society, Washington, DC. (1954), p 198.

[2] Handbook of Pharmaceutical Excipients, American Pharmaceutical Association, Washington, DC (1986).

[3] M.J. Akers, Canad. J. Pharm. Sci **11**, 1 (1976).

[4] F.A. Chrzanowski, L.A. Ulissi, B.J. Fegely, A.C. Newman, Drug Dev. Ind. Pharm. **12**, 783 (1986).

[5] A.L. Jacobs, Pharm. Manuf., **2**, 43 (1985).

[6] J.T. Carstensen, J.B. Johnson, W. Valentine, J.J. Vance, J. Pharm. Sci. **53**, 1050 (1964).

[7] D.A. Wadke and J. Jacobson, Pharmaceutical Dosage Forms, Volume 1, Marcel Dekker, Inc., New York (1980), H.A. Lieberman and L. Lachman, Eds., pp. 38–39.

[8] J.C. Callahan, G.W. Cleary, M. Elefant, G. Kaplan, T. Kensler and R.A. Nash, Drug Dev. Ind. Pharm., Vol. **8**, pp. 355 (1982).

[9] D.C. Monkhouse and L. Van Campen, Drug Dev. Ind. Pharm. **10**, 1175 (1984).

[10] K.J. Hartauer and J.K. Guillory, Drug Dev. Ind. Pharm. **17**, 617 (1991).

[11] M.J. Pikal and K.M. Dellerman, Int. J. Pharm. **57**, 22 (1986).

[12] L.D. Hansen, E.A. Lewis, D.J. Eatough, R.G. Bergstrom and D. DeGraft-Johnson, Pharm. Res. **6**, 20 (1989).

[13] M.L. Cotton, D.W. Wu and E.B. Vadas, Int. J. Pharm. **40**, 129 (1987).

[14] R.E. Gordon, C.L. Van Koevering and D.S. Reits, Int. J. Pharm. **21**, 99 (1984).

[15] L. Gu, R.G. Strickley, L.H. Chi and Z.T. Chowan, Pharm. Res. **7**, 379 (1990).

4 Prediction of the stability of drug products, new techniques and strategies

Klaus Dieter Gneuß, D-Biberach/Riß

1. Evaluation of stress test data A new strategy

1.1 Introduction

The chemical stability test is particularly important in the development of pharmaceutical products. A frequently used method is the accelerated isothermal test. Many methods for carrying out and evaluating isothermal stress tests have been described [1, 2, 3].

However, incorrect conclusions may have been drawn in some cases, particularly in the evaluation of the error of predicted values.

The way in which a prediction function C = f (t, T) is obtained is described below. This function makes it possible to minimize the bias as well as the error s (C). A five-step programme which shows how an optimal prediction function can be determined is described.

1.2 Theory

1.2.1 Determination of the order of the reaction

Conventional method. The first order reaction (1) is by far the most frequently encountered one in carrying out stress tests with pharmaceutical formulations.

$$C = 100 \cdot \exp(-k \cdot t) \qquad (1)$$

The logarithmic form of equation 1 is usually used for determining the rate constant $k(t^{-1})$:

$$\ln C = -k \cdot t + \ln 100 \qquad (2)$$

where k in equation 2 is generally determined by linear regression.

Use of the Weibull distribution for determining the order of reaction. Hancock and Sharp (4) used an equation for the analysis of solid-state reactions which is named after Avrami and Erofeev:

$$C = 100 \cdot \exp(-B \cdot t^{\beta}) \qquad (3)$$

Equation 3 is identical to the Weibull distribution, which is frequently referred to as the RRSBW distribution in the pharmaceutical technological literature.

In Bennett's notation, equation 3 becomes equation 4:

$$C = 100 \cdot \exp(-(t/t')^{\beta}) \tag{4}$$

C is the amount of substance not converted, in percent and t and t′ are times, t′ being the time in which 36.8% of the amount of substance considered have not been converted. β is a form factor.

The exponential distribution, i.e. the function for a first order reaction, is a special case of the Weibull equation (6), i.e. if $\beta = 1$, $k = 1/t'$

$$C = 100 \cdot \exp(-t/t') = 100 \cdot (-kt) \tag{5}$$

The parameters β and t′ are tested bei using reliability statistics (RRSBW); if $\beta = 1$, the reaction is a first order or pseudo first order reaction. In the logarithmic form of equation (4), β becomes a slope:

$$\ln\ln(100/C) = \beta \cdot \ln(t/t') \tag{6}$$

The reliability nets are built up in the form of equation (6). In the Weibull distribution (equation (4) or equation (6), (t/t′) models a characteristic time ratio which can be linked with predetermined orders of reaction.

Relative time ratios can be formed for all orders of reaction. For example, in order to produce the time ratio t1/t2 for a given first order reaction, equations 7a and 7b are first written:

$$k \cdot t1 = \ln(100/C1) \tag{7a}$$

$$k \cdot t2 = \ln(100/C2) \tag{7b}$$

Since the rate constant k is identical in both equations, equation 7c may be written as follows:

$$(1/t1) \cdot \ln(100/C1) = (1/t2) \cdot \ln(100/C2) \tag{7c}$$

From equation 7c, the relative time ratio (t1/t2) for a first order reaction is obtained as follows:

$$\frac{t1}{t2} = \frac{\ln(100/C1)}{\ln(100/C2)} \tag{7d}$$

Each relative time ratio can thus be written as a ratio of concentrations for a given order of reaction; the rate constant k need not be known.

Taking $t1 = t$ and $t2 = t'$, equation 8a is applicable for an order of reaction $n = 1$ in the Weibull coordinates:

$$\frac{t}{t'} = \frac{\ln(100/C)}{\ln(100/C')} \tag{8a}$$

and equation 8b is applicable for orders of reaction different from one

$$\frac{t}{t'} = \frac{((100/C)^{n-1})-1}{((100/C')^{n-1})-1} \tag{8b}$$

It is thus possible to derive the slopes of equation (6) which must result for a given reaction in the Weibull reliability net:
Equation (9) for n different from 1 and equation (10) for n = 1

$$\beta = \frac{((100/C)^{n-1})-1}{(n-1) \cdot \ln(100/C) \cdot (100/C)^{n-1}} \tag{9}$$

$$\beta = 1 \tag{10}$$

Fig. 1 shows the graphs for various orders of reaction using the coordinates $y = \ln\ln(100/C)$ and $x = \ln(t/t')$, and Table 1 summarizes the corresponding slopes of these graphs for selected concentrations.

Hancock et al. (4) in 1972 and Okusa (7) in 1975 used the Weibull distribution or the Avrami equation for determining time laws. They used the concentration range of 15%–50% of reacted starting substance for determining orders of reaction.

Thus, the RRSBW distribution can be used for discriminating between orders of reaction. This fact is also used in the kinetic consideration of the dissolution or release of active substances from pharmaceutical products [8, 9, 10].

The Weibull reliability net is particularly suitable for visualizing stress test

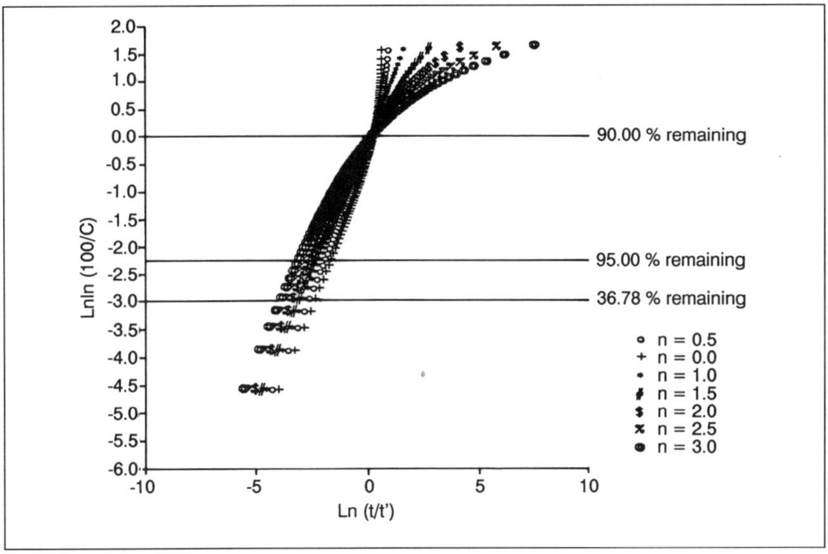

Fig. 1: Graphs for orders of reaction between 0 and 3 in the Weibull reliability net

Table 1: Summary of the slopes of graphs in the Weibull reliability net for orders of reaction from 0 to 3 in the concentration range from 1% to 99%

Concentration	Order of reaction			
C (%)	0.0	1.0	2.0	3.0
99	1.0050	1.0000	0.9945	0.9900
95	1.0261	1.0000	0.9748	0.9504
90	1.0546	1.0000	0.9491	0.9017
80	1.1204	1.0000	0.8963	0.8067
50	1.4427	1.0000	0.7213	0.5410
20	2.4853	1.0000	0.4971	0.2982
10	3.9087	1.0000	0.3909	0.2150
5	6.3424	1.0000	0.3171	0.1665
1	21.4976	1.0000	0.2150	0.1086

results since the isotherms can be tested for parallelism in one picture (in this context, see Section 3).

1.2.2 Checking the agreement of the Arrhenius parameters within the isotherms of the stress test

The Arrhenius equation describes the temperature dependence of the rate constant k for a given order of reaction

$$k = A \cdot \exp(-E/RT) \qquad (11)$$

A is the preexponential factor and E is the Arrhenius activation energy, T is the absoute temperature and R is the gas constant. If a chemical reaction in the temperature range invetigated strictly obeys an Arrhenius equation, the parameters A and E are exactly the same at all temperature levels investigated; A and E are temperature-independent.

Under these conditions, the following is true:

$$A = k1 \cdot \exp(-E/RT1) = k2 \cdot \exp(-E/RT2) \qquad (12)$$

If equation 12 is solved to obtain E, the following is true:

$$E = R \cdot \ln(k2/k1) \cdot (T1T2/(T2\text{-}T1)) \qquad (13)$$

Equation 13 makes it possible to link every ith k value of an ith temperature with every jth k value of a jth temperature and to calculate the corresponding activation energy E.

For example, if stress tests have been carried out at 5 temperature levels and the amounts of active substance have been analyzed at 5 different times within the temperature, it is possible to calculate 250 values of the activation energy.

These calculated E values are scattered about a mean value E and, for example, temperature errors produce marked contrasts. The calculated activation energies E (k1, k2, T1, T2) are advantageously visualized for interpretation. If calculated activation energies are represented, for example, in the form of a histogram, it is easy to see which isotherms belong to the same population (see the statements in Section 3.3).

1.2.3 Determination of the parameters of a prediction function

If a chemical reaction takes place as a first order reaction in the experimentally tested range, then combining equation 5 and equation 11 gives the following for a prediction function:

$$C = 100 \cdot \exp(-t \cdot A \cdot \exp(-E/R \cdot T)) \qquad (14)$$

Equation 11 is usually expressed in logarithmic form (11a) for the determination of the parameters A and E. The parameters were then calculated by linear regression.

$$\ln k = \ln A - E/R \cdot T \qquad (11a)$$

The Arrhenius parameters can be determined directly from equation 11 by the method of non-linear regression, and the standard deviations of the calculated mean values of these parameters, which are required for analyzing the errors in C, are also obtained.

The advantages of non-linear regression for the statistical prediction of the stability of active substances compared with linear regression after transformation are discussed in detail by King, Kung and Fung [11].

Ebel [12, 19, 25] reports on the danger of falsification of results when using linear regression after transformation compared with non-linear regressions. In this paper, the parameters were calculated by non-linear regression with the aid of the multidimensional Newton method according to Ebert and Ederer [13].

1.2.4 Calculation of the error of a predicted concentration C

The error of a predicted concentration can be estimated using Gauß's error propagation law (equation 15) [14, 15].

$$s^2(C) = (\partial C/\partial A)^2 \cdot s^2(A) + (\partial C/\partial E)^2 \cdot s^2(E) \qquad (15)$$

The partial derivatives $\partial C/\partial A$ and $\partial C/\partial E$ of the prediction equation (14) are as follows:

$$\partial C/\partial A = -100 \cdot t \cdot \exp(-E/R \cdot T) \cdot \exp(-t \cdot A \cdot \exp(-E/R \cdot T)) \qquad (16)$$

$$\partial C/\partial E = (100 \cdot t \cdot A/R \cdot T) \cdot \exp(-E/R \cdot T) \cdot \exp(-t \cdot A \cdot \exp(-E/R \cdot T)) \qquad (17)$$

Table 2: Results of a stress test by Kottke[16] with tolazoline (Series 1)

Stresstemperature (°C)	Storage time (h)	Concentration (%)	k·1000 (h^{-1})
90	6	95.63	7.45
	24	83.51	7.50
	48	69.68	7.53
	72	58.83	7.37
	96	48.19	7.39
80	12	95.48	3.85
	24	91.63	3.64
	48	83.58	3.74
	72	75.76	3.86
	96	72.51	3.35
70	24	94.85	2.20
	48	92.11	1.71
	72	85.63	2.15
	96	81.53	2.13
	120	77.63	2.11
60	48	94.00	1.29
	96	86.32	1.53
	144	82.79	1.31
	192	75.73	1.45
	240	73.97	1.26
50	48	97.82	0.46
	96	95.51	0.48
	144	93.23	0.49
	192	91.96	0.46
	240	88.33	0.52

1.3 Procedure

To demonstrate the novel procedure for the evaluation of stress tests, published results by Kottke [16] were used with the kind permission of the author. In addition to stress test results, this paper also contains an evaluation after storage over the predicted period, providing a good check. Kottke's values are summarized in Table 2.

1.3.1 First step: Visualization of the stress test results in the RRSBW net

In the RRSBW net, it is possible to check very simply and rapidly whether the curves obtained when lnln(100/C) is plotted against ln(t) have the same form, i.e. order of reaction, at all temperatures tested. Furthermore, it is possible to

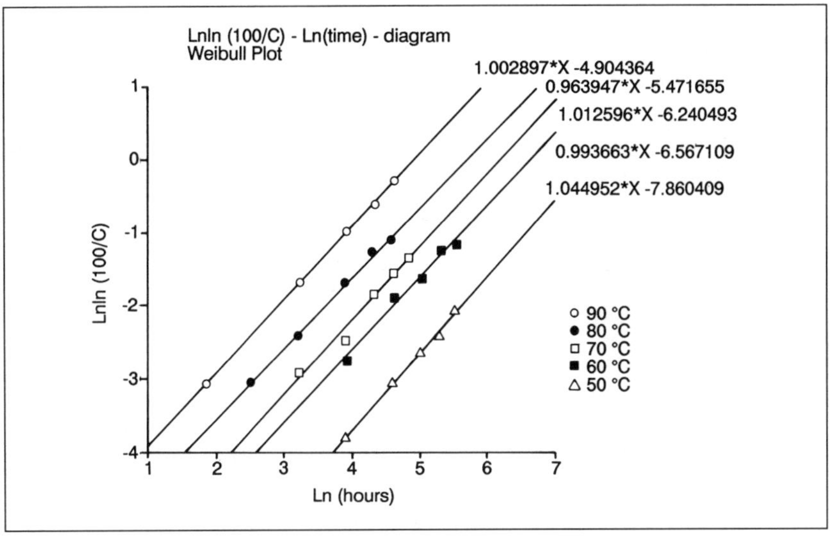

Fig. 2: Plot of the values from Table 2 in the RRSBW reliability net

recognize from the family of curves whether the distances between the curves are approximately the same, i. e. whether Arrhenius parameters showing good agreement are expected. The values in Table 2 are shown in the RRSBW reliability net in Fig. 2.

Evaluation: A striking feature is that the straight line obtained when the values of lnln(100/C) are plotted against ln(t) for storage at 60 °C is closer to the 70 °C isotherm than to the 50 °C isotherm. The isotherms appear to have approximately the same slopes.

1.3.2 Second step: Non-linear regression using the Weibull distribution as a model

With the Weibull distribution (equation 4) as a model, the parameters summarized in Table 3 are obtained by the non-linear regression:

If β is distributed according to $N(\mu, \sigma^2/n)$, an empirical mean value can be compared with the mean value of a population with a normal distribution [17].

$$t = \frac{x - \mu_o}{s} \cdot \sqrt{n} \qquad (18)$$

The question posed is whether the zero hypothesis $\mu = \mu_O = 1$ is accepted with a statistical certainty of s = 99% (Table 4).

Evaluation: With a statistical certainty of s = 99%, the zero hypothesis $\mu = \mu_O = 1$ is not rejected for any isotherm. Hence, a first order reaction can be calculated using the model.

1.3.3 Third step: Checking the agreement of the Arrhenius Parameters A and E by properly designed experiment

Since β in the Weibull function does not differ significantly from 1 at all temperatures tested, a first order reaction is expected (see equation 5).

$$C = 100 \cdot \exp(-kt)$$

When this equation is solved for k, the following is obtained:

$$k = (1/t) \cdot \ln(100/C) \qquad (19)$$

Using equation 19, the rate constant k is calculated from each pair of concentration-time values in Table 2 (see table 2).

Activation energies E are now calculated using equation 13 and the k values from Table 5. Each k value of a stress temperature is combined with all other k values of the remaing stress temperatures.

The rounded, integral E values are summarized in Table 5.

The E values listed in Table 5 are plotted as a function of the nth calculation (time plot) for visualization in Fig. 3.

In Fig. 4, the calculated E values from Table 5 are shown as a histogram.

Table 3: Summary of the calculated parameters t' and β of the Weibull function and the standard deviations s(β) and s(t') of the data from Table 2

T (°C)	t' (h)	s (t')	β	s (β)	$\Sigma\,(\gamma\text{-}\gamma\text{mean})^2$
90	132.61	1.30	1.0072	0.0123	0.5215
80	313.77	26.83	0.9204	0.0500	3.8472
70	427.22	39.63	1.0734	0.0634	2.5083
60	848.82	107.26	0.9133	0.0712	5.9298
50	1702.07	215.45	1.0820	0.0616	0.5231

Table 4: Significance test for the differences in the form factor $\beta^{a)}$

Storage temperature (°C)	t = tabulated value S = 99%$^{b)}$	t-value	Zero hypotesis accepted
90	5.841	1.3089	yes
80	5.841	3.5598	yes
70	5.841	2.5888	yes
60	5.841	2.7229	yes
50	5.841	2.9766	yes

a) Zero hypothesis: ß = 1 (Table 3)
b) Degrees of freedom = 3 for all temperatures, n = 5

Table 5: Summary of all calculated E values

Calculated activation energy E (kJ/mol)																					
1th condition		2nd condition																			
T1/k1 °C	T2/°C k2	80					70					60					50				
		1	2	3	4	5	1	2	3	4	5	1	2	3	4	5	1	2	3	4	5
90	1	70	76	74	70	85	63	76	64	65	65	59	53	58	55	60	68	67	67	68	65
	2	71	77	74	71	86	64	77	65	65	66	59	53	59	55	60	68	67	67	68	65
	3	71	77	75	71	86	64	77	65	65	66	59	53	59	55	60	68	67	67	68	65
	4	69	75	72	69	84	63	76	64	64	65	58	53	58	55	59	68	67	66	68	65
	5	72	78	76	72	87	64	77	65	66	66	60	54	59	56	60	68	67	67	69	66
80	1						56	82	59	60	61	54	45	53	48	55	67	66	65	67	64
	2						51	76	53	54	55	51	42	50	45	52	65	64	64	66	62
	3						53	79	55	57	58	52	44	51	46	53	66	65	64	66	63
	4						56	82	59	60	61	54	45	53	48	55	67	66	65	67	64
	5						42	68	44	46	47	47	38	46	41	48	63	62	61	63	59
70	1											51	35	49	40	53	72	70	70	73	67
	2											27	11	25	16	29	61	59	58	61	55
	3											49	32	47	38	51	71	69	69	71	66
	4											48	31	46	37	50	71	69	68	71	65
	5											47	30	45	36	49	70	68	68	71	65
60	1																92	89	87	93	82
	2																108	104	103	108	97
	3																94	90	89	94	83
	4																103	99	98	103	92
	5																90	86	85	91	79

Starting from a population with a normal distribution, i.e. N (μ, σ^2/n), equation 20 gives the following for the confidence interval of the mean value with a certainty of s = 99%:

$$\text{xmean} \pm t \cdot s/\sqrt{n} = (63.83 \pm 2.54)\ \text{kJ mol}^{-1} \qquad (20)$$

If the E values are considered in detail, it is found that the extreme values of the histogram (Fig. 4) originate from values which are calculated using rate constants from the group of the 60 °C isotherms. If they are calculated using 70 °C rate constants, the E values are extremely low. If the same 60 °C rate constants are replaced with the 50 °C rate constants, the corresponding E values are extremely high.

Figure 3.1 summarizes the E values as a function of the nth combination, only 50 °C, 70 °C and 90 °C results being used.

This contrast emerges in the RRSBW net (Fig. 2) itself; there, the 60 °C isotherm is closer to the 70 °C isotherm than to the 50 °C isotherm. A systematic error must be present here.

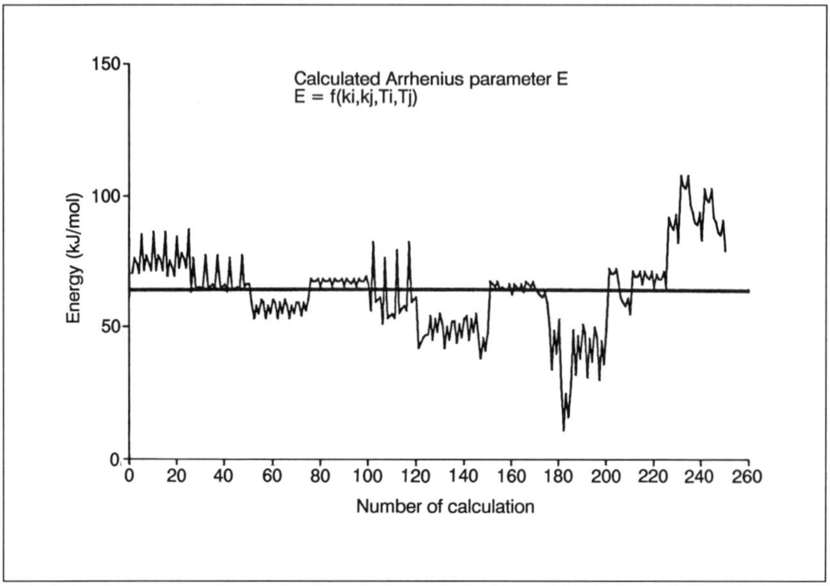

Fig. 3: Dependence of the calculated activation energy E on the nth combination of k1 with k2

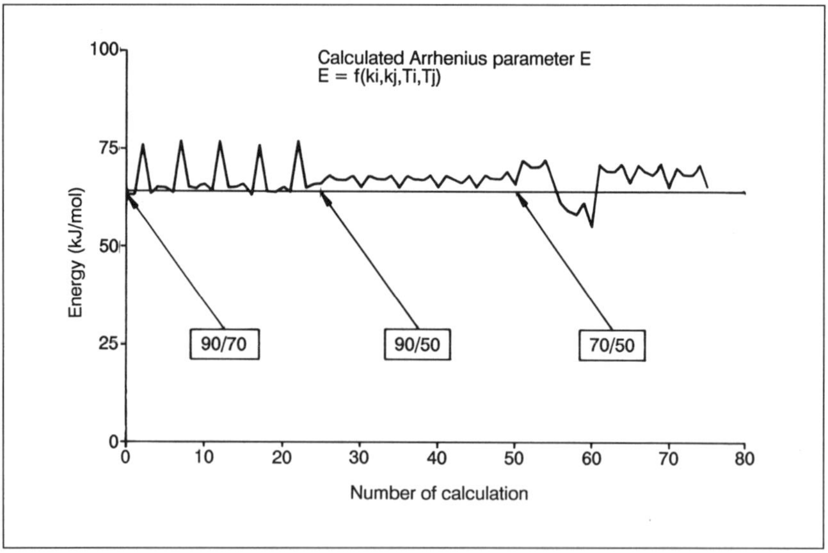

Fig. 3.1: Summary of the E values as a function of the nth combination, only 50 °C, 70 °C and 90 °C results being used

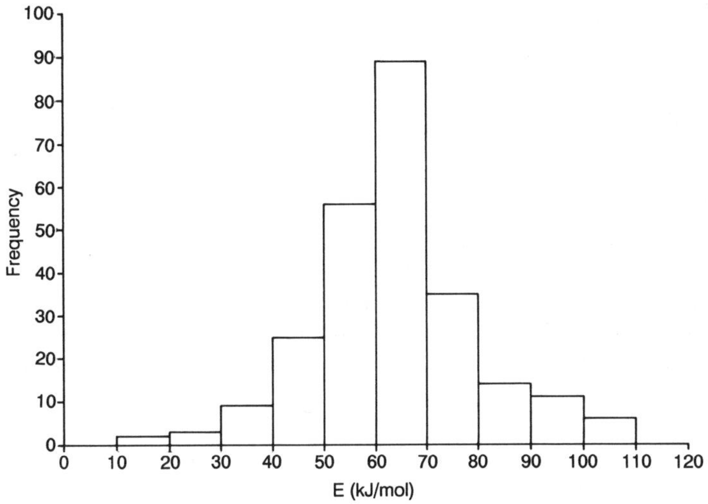

Fig. 4: Histogram of the E values (frequency distribution) of Table 5 with the class width of 10 kJ · mol^{-1}; x = 63.83 (kJ mol^{-1}); s = 15.5 (kJ · mol^{-1}); n = 250

A contrast is observed in the calculated E values even when 80 °C rate constants are used. If the 80 °C rate constants are replaced with the 90 °C rate constants, the resulting E values are considered to be higher than the calculated mean value. However, if they are replaced by the 70 °C rate constants, the E values obtained are considerably below the calculated mean value.

On the basis of these results, it must be assumed that the 80 °C isotherm (like the 60 °C isotherm) contains a systematic error.

If, on the other hand, the calculated E values are examined using the rate constants of the isotherms of 90 °C and 70 °C, 90 °C and 50 °C, 70 °C and 50 °C, it is found they occur in the most frequently occupied class between 60 kJ · mol^{-1} and 70 kJ · mol^{-1} (see Fig. 4). The E values which are outside this class are exclusively those which were combined with the 48 h value of the 70 °C isotherms.

Thus, it can be seen that three of the five stress temperatures have largely identical Arrhenius parameters.

These isotherms are further used for minimizing the error of prediction.

1.3.4 Fourth step: Calculation of the Arrhenius parameters A and by non-linear regression and formulation of the prediction equation

The Arrhenius parameters A and E are determined by non-linear regression using equation 11:

$$k = A \cdot \exp(-E/RT)$$

A concentration after storage for one year (= 8760 h) at 25 °C (= 298.2 K) is then predicted using prediction equation 14.

Table 6 summarizes the parameters used and the calculated parameters.

The following prediction equation is obtained using the Arrhenius parameters A and E listed in Table 6:

$$C = 100 \cdot \exp(-18214738.8 \cdot t \cdot \exp(-65246.0/R \cdot T))$$

For t = 8760 h, T = 298.2 K and R = 8.3143(J/K · mol), a concentration of 55.64% of active substance is predicted (actual content according to Kottke [16]): 55.63%).

Evaluation: The actual content after storage is excellently predicted using the selected stress test data.

1.3.5 Fifth step: Calculation of an error of a predicted concentration C

The error in a predicted concentration s(C) is calculated using Gauß's error propagation law (see equations 15, 16 and 17).

Table 6: Selected data and Arrhenius parameters A and E calculated therefrom by non-linear regressions, and errors s (A) and s (E)

Parameters used		Parameters calculated			
Storage Temperature (°C)	Number of k values[a)]	A (1/h)	s (A) (1/h)	E (J/mol)	s (E) (J/mol)
90	5; all values	18214738.8	1370266.1	65246.0	226.2
70	4; all values except for 48 h value				
50	5; all values				

a) From Table 2

Table 7: Data sets and calculated Arrhenius parameters with the standard deviations s (A) and (E)

No.	Parameter used: Storage temperature (°C)	Number of k values[a)]	Parameters calculated: A (1/h)	s (A) (1/h)	E (J/mol)	s (E) (J/mol)
1[b)]	90	5	18214738.8	1370266.1	65246.0	226.2
	70	4				
	50	5				
2	90	5	20740582.8	5545129.6	65683.2	798.4
	80	5				
	70	5				
	60	5				
	50	5				
3	90	5	82544011.0	16481982.9	69814.1	597.5
	80	5				
	70	5				
	50	5				
4	90	5	33969702.8	5469778.3	67110.5	483.9
	70	5				
	50	5				

a) Selected values from Table 2
b) Regarded as belonging together and hence optimal

The following error s(C) is calculated from the data in Table 6: s(C): 3.86%. Number of values: 14, degrees of freedom: 12
assuming a population with a normal distribution, $N(\mu, \sigma^2/n)$, it is possible to calculate a confidence interval for the predicted mean value with a certainty of S = 95%:

$$55.64\% \pm 2.18 \cdot 3.88\%/\sqrt{14} = 55.64\% \pm 2.26\%$$

A predicted concentration C is between 53.39% and 57.89% with a certainty of 95%.

1.4 Comparison of predicted results observed with various data sets

To demonstrate that both very good predictions of an expected value are obtained and minimization of the error is achieved by checking the agreement of the Arrhenius parameters A and E by a suitable test method, further data sets from Table 2 were used for predictions. Table 7 summarizes the data used and the resulting Arrhenius parameters A and E and the errors s(A) and s(E).

The data set regarded as optimal in this work is listed under No. 1.

Table 8: Summary of the concentrations C predicted using the Arrhenius parameters of Table 7, of the standard deviations s (C) and of the confidence intervals with a certainty of 95%

No.	Number of values	Predicted Concentration C (%)	Standard-deviation s (C)	Confidence interval of the predicted c (%) with S = 95%
1[b)]	14	55.64	3.86	53.39 – 57.89
2	25	57.54	12.36	52.42 – 62.66
3	20	65.67	8.85	61.61 – 69.73
4	15	59.76	7.79	55.42 – 64.10
5[c)]	Kottke (16)[b)]	54.00	–	50.29 – 57.56
6	Storage experiment Kottke[16)]: (experiment: 1 year, 25 °C) = 55.63%			

a) Values from Table 7
b) Optimized result
c) Determined by a weighting method

Table 8 summarizes concentrations predicted using the parameters of Table 7, and the corresponding errors with the confidence intervals.

Figure 5 shows the three-dimensional C (t, T) plot for the predicted function 1 in Table 7.

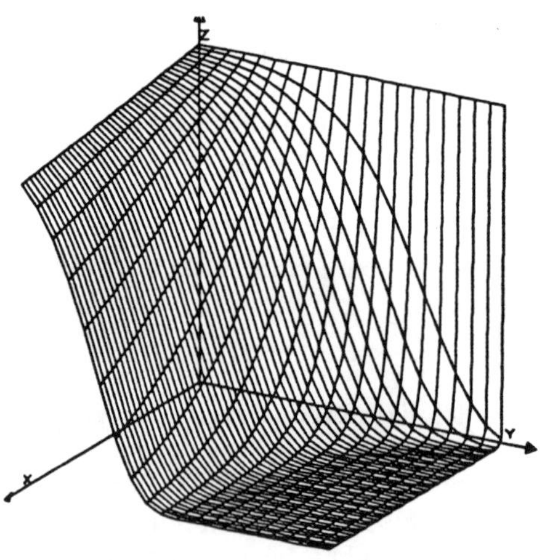

Fig. 5: The three-dimensional C (t, T) plot for the predicted function 1 in Table 7. Z = Concentration C/%, X = Time/h, Y = Temperature/K

1.5 Discussion

In pharmaceutical formulations, active substances do not generally have unlimited chemical stability. If it is desired to obtain information about the time and temperature dependence of a decomposition reaction, stress tests are carried out.

With the method used in this paper, i.e. the non-linear calculation of parameters and subsequent error calculation in a predicted value by the error propagation law, the incorrect step employed in the "classical evaluation method" is avoided, namely the calculation of errors after transformation and subsequent linear regression.

Another advantage of the new strategy for evaluation is that both the bias of a predicted concentration and the variance for this value are minimized.

All decomposition reactions which generate the slope 1 in the Weibull reliability net in the test period can be reliably treated by the proposed method of evaluation. These reactions are:

- Genuine first order reactions
- Pseudo first order reactions
- All orders of reaction < 1 and > 1 if the decrease in content is $< 10\%$ (see Fig. 1 and Table 1)

Furthermore, complicated reactions whose isotherms have identical slopes differing from 1 in the Weibull reliability net can be evaluated in an analogous manner. The evaluation of this type of reaction forms the subject of a planned paper. In these cases, $(1/t')^{\beta}$ is the temperature-dependent parameter. Further data analysis must then show whether the Arrhenius parameters A and E are temperature-dependent.

Equilibrium reaction and catalytic reactions require considerable theoretical and experimental work. However, the category in which the relevant reaction has to be classed can be determined in this case too from the shape of the isotherms or the distance between the isotherms, using the Weibull reliability net.

Fortunately, by far the most decomposition reactions of active substances of pharmaceutical formulations are formally of the first order type and the Arrhenius parameters A and E are temperature-independent, so that the proposed strategy can successfully be used for evaluating stress tests.

2. The linearized form of the Arrhenius equation and the problem of error determination for $k = k(T)$ from $\ln k = \ln k(1/T)$

It is anachronistic that even today stress test results are evaluated via $\ln k = \ln k(1/T)$ when there are computer programs which enable the non-linear regression analyses to be carried out directly in the model. We use, for exam-

Table 9: Parameter Table

0	1 Param.	2 Final values	3 Std Errors	4 T value	5 Sig. Lev.	6 Dep. values
1	A	31244815.129592	13163545.632324	2.374	0.0337	1
2	E	66882.557791	1266.112961	52.830	0.0001	1

ple, the software package RS/1 from BBN [23]. Table 9 shows the calculated parameters A and E with the corresponding errors s(A) and s(E) for any stress test evaluation.

It is now possible to use error propagation calculations to calculate errors in predicted rate constants k, without any transformation operations. If nevertheless the detour via the linearized form of the Arrhenius equation is taken and the upper and lower confidence limits of lnk are calculated from the confidence intervals, k (upper) and k (lower) are determined from these and furthermore a concentration of an active ingredient C(t,T) is predicted, then the following questions must be asked:

Is the predicted value determined on a linear basis identical to that from a nonlinear calculation?

Answer: The values may not be identical since, for example, the formulae cannot be identical when the least squares method is used. Numerical differences in the result are the consequence, as can be seen from Fig. 6.

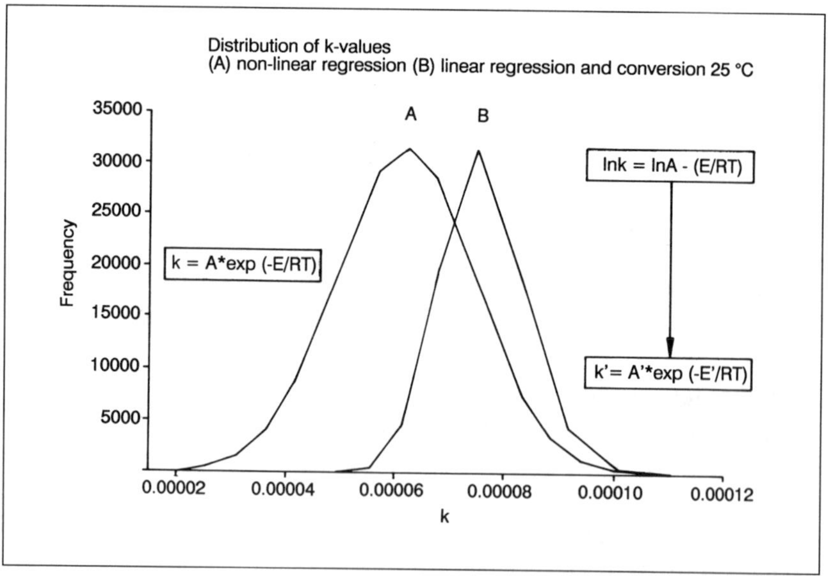

Fig. 6: The results of linear and nonlinear regression for any stress test evaluation

Are the calculated errors of the two methods identical?

Answer: The errors need not be identical. It would in any case be necessary to check whether the variance of the nonlinear form is produced via the linearized form. If the variances are not identical, the homoscedasticity principle is violated, as reported by Ebel et al. [19]).

Fig. 6 shows that the k values calculated via nonlinear and via linear regression and retransformation are scattered to different extents about different mean values at 25 °C. The differences determined may be very far-reaching.

The confidence intervals have a normal distribution about a predicted value lnk = lnk(1/T). Is this normal distribution retained after retransformation?

Answer: This requirement cannot be applicable for lnk and k, and lnk and k cannot simultaneously have a normal distribution about the same mean value (Fig. 6).

There are no good reasons for continuing to use the linearized form of the Arrhenius equation.

3 Long-term stability test

3.1 Cost of analysis with known precision of the method – An approximation –

Before a sample, for example from the long-term test, is analysed, a careful and conscientious validation of the method is generally carried out using the finished product.

During development of the method, the precision is determined in addition to the accuracy; usually, 10 individual analyses are performed.

Three fictitious standard deviations are considered in detail below: s = 1, 2 and 3 (units of the parameter) with a mean value of 100 units. Table 10 summarizes the confidence limits of the variances and of the standard deviations with 95% certainty. The test is carried out using the chisquared distribution according to Pfanzagel [23].

Table 10: Summary of fictitious variances or standard deviations and their confidence limits with 95% certainty

Fictitious scatter		Confidence limits of the scatter with 95% certainty			
SDEV	VAR	Lower limit		Upper limit	
		VAR	SDEV	VAR	SDEV
1	1	0.47	0.69	4.28	2.07
2	4	1.89	1.37	17.14	4.14
3	9	4.26	2.06	38.57	6.21

Table 11: Fictitious scatters and the corresponding median values (n = 10)

Random sample		Median estimates	
SDEV	VAR	SDEV	VAR
1	1	1.04	1.08
2	4	2.08	4.34
3	9	3.14	9.74

If the variances exhibited scatter in the stability test to the extent shown in Table 10, no reliable analyses would be possible since a substantiated standard deviation of the method of s = 1 might also be 2.

But how is it that reliable analyses can nevertheless be carried out in practice?

This is due to the fact that analytical results of validated methods are unbiased.

Sachs [17] states that "... statistics such as mean values must be unbiased, i.e. in the case of repetitions it must be possible to demonstrate that the mean value of the estimates for finite n is centred around the true value in the population."

In analysis, a bias often occurs only because of systematic errors.

Pfanzagel [23] uses the median variance (see Table 11) for the unbiased variance.

Unbiased results can be particularly readily simulated using normal distributed random numbers.

3.1.1 Simulation of stability tests with normal distributed random numbers

To generate random samples which originate from the population N(μ, σ), an algorithm from Bruhn [21] is used. Table 12 summarizes the results of test runs.

The simulation plan is as follows:

Three analyses in each case at the beginning and one analysis in each case after 3, 6, 12, 24, 36, 48 and 60 months.

Two questions were of particular interest:

1. What was the trend in the confidence intervals when there was no decrease in content?

Table 12: Summary of the given and obtained parameters μ and σ

Number of random numbers generated	Mean values		Standard deviation	
	Given	Obtained	Given	Obtained
5000	100.00	99.99	1.0	0.9995
5000	100.00	99.98	2.0	1.9990
5000	100.00	99.97	3.0	2.9985

Table 13: Summary of the predicated values and their lower confidence limits with 95% certainty after storage for 60 months

Condition	SDEV of the population	Concentration after 60 months/%	
		Predicted	Lower confidence limit 95% certainty
100 % = constant	1	100.0	98.3
	2	101.1	97.8
	3	99.5	93.9
from 100% to 95% in 60 months	1	94.8	93.1
	2	92.8	91.2
	3	93.7	88.6

2. What is the trend in the confidence intervals when the true content decreases during storage from 100% at the beginning to 95% after 60 months?

The procedure for calculating the regression corresponds to the FDA Guidelines [20, 22].

3.2 Summary

On the basis of the results obtained by simulation, it has been shown that unbiased values – including analytical results – lead to good results with the proposed test plan. In other words, it is also possible to carry out single analyses if the scatter of the method is known. Where the results are outside the tolerance limits (for example outside the 2 s limit of the error of the method), multiple analyses are in any case immediately indicated.

References

[1] Thoma, K., "Arzneimittelstabilität", Frankfurt, 1978

[2] Grimm, W., Schepky, G., "Stabilitätsprüfung in der Pharmazie", Editio Cantor, Aulendorf, 1980

[3] Kersten, D., Döber, B., Pharmazie **3a**, 213–225 (1983)

[4] Hancock, J. E., Sharp, K. H., J. Am. Ceram. Soc. **55**, 74–77 (1972)

[5] Müller, P. H. (Editor), "Wahrscheinlichkeitsrechnung und mathematische Statistik, Lexikon der Stochastik", Berlin 1975, 2nd ed., P. 321

[6] Adolphi, G. (competent author), "Lehrbuch der chemischen Verfahrenstechnik", Leipzig, 1969, 2nd ed., pp. 514–539

[7] Okusa, N., Chem. Pharm. Bull., **23**, 794–802 (1975)

[8] Langenbucher, F., J. Pharm. Pharmacol. **24**, 979–981 (1972)

[9] Christensen, F. N., Hansen, F. Y., Bechgaard, H. J., J. Pharm. Pharmacol., **32**, 580–582 (1980)

[10] Heinrich, J., Chalabala, M., Rak, J., Acta Pharm. Technol., **32**, 94–96 (1986)

[11] King, S.-Y. P., Kung, M.-S., Fung, H.-L., J. Pharm. Sci. **73**, 657–662 (1984)
[12] Ebel, S., GDCH-course-papers: "Kalibrierung und Validierung in der analytischen Chemie", Würzburg, 1986
[13] Ebert, K., Ederer, H., "Computeranwendung in der Chemie", 2nd ed., VCH-Verlag, Weinheim 1984, pp. 321–326
[14] Dörfel, K., Statistik in der analytischen Chemie, 3rd ed., VCH-Verlag, Weinheim 1984, pp. 51–72
[15] Frauenhofer, P., Huber, P., "Einführung in die Physik", Ernst Reinhardt Verlag, Basel 1964, vol. 1, 2nd ed., pp. 37–42
[16] Kottke, D., Zbl. Pharm. **117**, 918–923 (1978)
[17] Sachs, L., Angewandte Statistik, 5th ed., Springer-Verlag, Berlin, Heidelberg, New York 1978, pp. 195–197
[18] Gneuß, K.D., Acta Pharm. Technol. **35**, 67–74 (1989)
[19] Ebel, S., Fleischer, P., Ledermann, M., Wüstenhagen, U., Acta Pharm. Technol. **35**, 210–217 (1990)
[20] Food and Drug Administration, Guideline for Submitting Documentation for the Stability of Human Drugs and Biologics", 1987, Rockville, MD
[21] Bruhn, J., "Statistische Verfahren", 1st ed., Vieweg Verlag, Braunschweig, 1986, pp. 91–94
[22] Carstensen, J. T., "Drug Stability", 1st ed., Marcel Dekker, New York, 1990
[23] Pfanzagel, J., "Allgemeine Methodenlehre", 4th ed., vol. 2, Walter de Gruyter Verlag, Berlin, New York, 1983, pp. 72–75, 195–197
[24] RS/1® User's Guide
BBN Software Products Corporation, 1987
10 Fawcett Street, Cambridge, MA 02238
[25] Ebel, S., Ledermann, M., Reyer, B., Eur. J. Pharm. Biopharm. **37**, 80–87 (1991)

Acknowledgements
I would like to thank Dr. Grimm for fruitful discussions and Mr. Friedel for the careful and conscientious work on the computer.

5 A critical review of high-performance liquid chromatography, supercritical fluid chromatography and capillary zone electrophoresis in drug analysis

W. Steuer, D-Ingelheim

1. Introduction

High-performance liquid chromatography (HPLC) is a well established method for the purity control and assay of drugs in galenical forms both in the release analysis and in stability investigations or in biological matrices. However the use of supercritical fluid chromatography (SFC) has greatly increased. At present, in addition to non-polar compounds, polar substances and large molecules can be separated. SFC [1] has been extended to include applications involving ionic compounds. During the last few years, capillary zone electrophoresis (CZE), was developed mainly for the separation of proteins. There has been comparatively little interest in the analysis of pharmaceuticals.

This paper compares the advantages and limitations of HPLC, SFC and CZE in drug analysis. Most of the example separations using SFC and CZE were carried out using laboratory-constructed apparatus, which has been described elsewhere [1, 2].

Drug analysis can be divided into the following subgroups: drug substance analysis, stability investigations of galenical forms, analysis of excipients, determination of drugs in biological fluids and biotechnological analysis. Each of these subgroups possesses an additional profile of requirements in terms of efficiency, performance, selectivity, sensitivity and sample preparation, which must be taken into account when deciding on the suitability of a separation method.

2. Evaluation criteria

This comparative study is based on the following evaluation criteria: efficiency (N); performance (N/t) (N = number of plates; t = time); sensitivity; parameters available for modification of selectivity; method development time; sample preparation; and orthogonality of information.

Table 1: Analysis times required for HPLC, SFC and CZE

No. of components	N	Times required (s)			
		HPLC (20°C)	HPLC (90°C)	SFC	CZE
5	1000	4.2	1.1	0.8	–
10	10000	56	15.2	10.8	–
20	50000	480	140	95	60
50	500000	32000	10300	7100	600

2.1 Efficiency and performance

In many analytical problems a high number of plates is desirable as demonstrated elsewhere [3, 4, 5]. Table 1 compares the necessary analysis times in HPLC, SFC and CZE for the separation of a given number of species with a given number of plates.

The calculations made by Steuer et al. [6] are not explained in greater detail here.

Table 1 demonstrates however that HPLC should be used when a small number of components have to be separated in a short time. This is a typical requirement for the stability analysis of galenical forms where a large number of samples usually have to be processed overnight. In addition to sufficiently short analysis times, reliable and automatic systems are now also required.

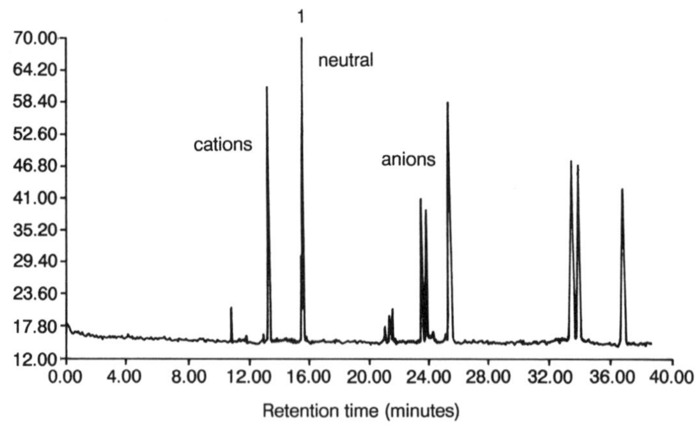

Fig. 1: Analysis of a stressed pharmaceutical alkaloid (1) by CZE. Applied voltage, 25 kV; buffer, water-acentonitrile (60:40) – 30 mM Na_2HPO_4; capillary, 90 cm × 50 μm I.D.

When a high number of plates is required, SFC and CZE are to be preferred. For the determination of drug purity, where often 10–15 peaks have to be separated, in some instances SFC may be a useful alternative to HPLC whereby it should be borne in mind that when using gradient elution in the Reversed Phase HPLC sufficiently high peak capacities are often possible with fewer technical resources.

CZE is the most appropriate technique for the separation of very complex mixtures, where more than 100000 plates are required (Fig. 1). Typical examples are the purity control of drugs and biotechnological analysis such as the separation of protein digests.

2.2 Sensitivity

The sensitivity of a separation method depends on both the loading capacity of the system and the nature of the available detectors. It must be borne in mind that in day-to-day practice, concentration sensitivity is the crucial factor and not mass sensitivity as frequently propagated in conjunction with microsystems [7].

In contrast to HPLC and SFC, CZE can only be used in a miniaturized form, resulting in constraints in terms of loading capacity and sensitivity. It should be emphasized that for absorbance and fluorescence detection it is the concentration sensitivity and not the minimum detectable amount that is relevant in drug analysis.

The latter can often be deceptive in capillary SFC and CZE. When working with packed conventional columns, UV detection in SFC and HPLC is comparable in terms of sensitivity and loading capacity. The noise level may be greater in SFC owing to density-related refractive index changes in the highly compressible supercritical fluid.

Practical experience shows that in the routine work the concentration sensitivity in HPLC is at least ten times better than that in CZE (Fig. 2).

In both instances the same sample concentration was injected (1 mg/ml). The special Z-celle described by Chervet [8] may allow an increase in UV-concentration sensitivity.

The poor sensitivity of CZE cannot be totally compensated for by an increase in the amount injected. Because the efficiency was observed to decrease by a factor of four following a 100-fold increase in the sample concentration (Fig. 3). This leads to an unacceptable loss of efficiency and resolution. Additionally, the signal-to-noise ratio does not increase linearly with the amount injected because of the additional peak broadening, which results in a less than linear increase in peak height. Thus in addition to the optimization of selectivity, the amount injected also has to be optimized, subject to resolution requirements [6].

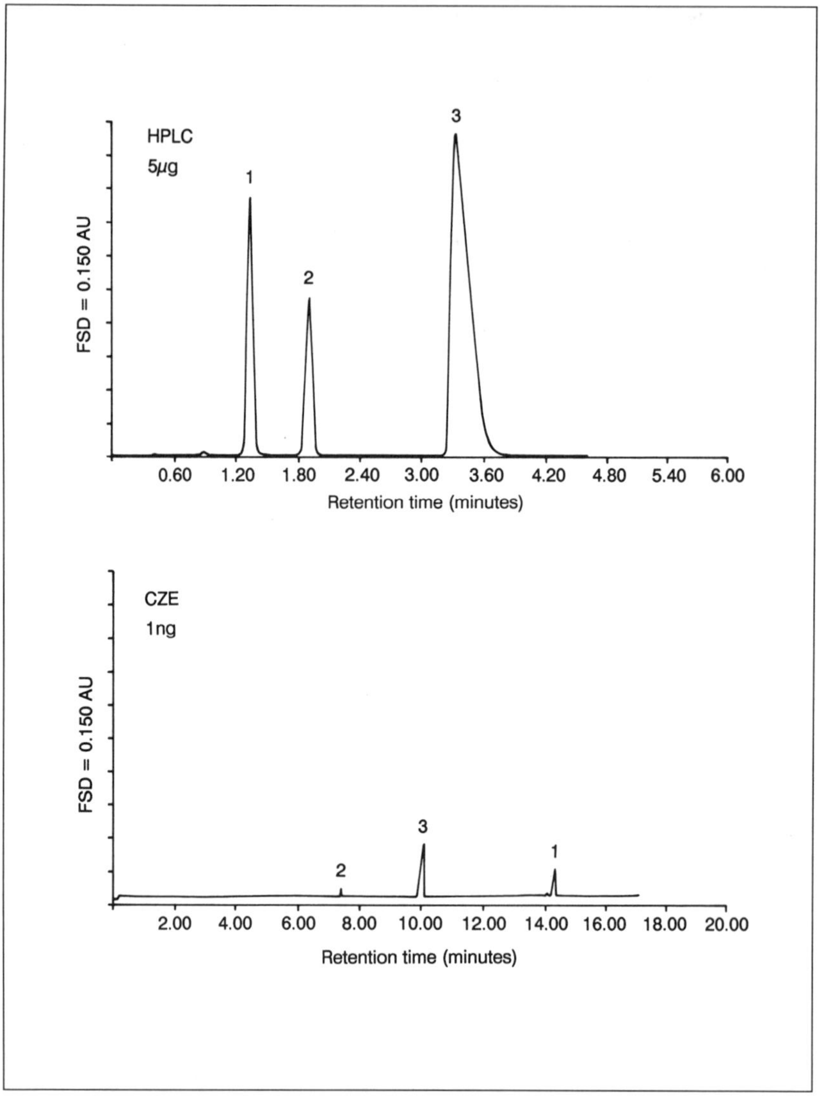

Fig. 2: Sensitivity comparison for HPLC and CZE; separation of spirapril and its degradation products. HPLC: column, Brownlee RP 18,5 μm, 100 mm × 4.6 mm I. D.; eluent, acetonitrile-water (50:50) – 0.5 mM tetramethylammonium hydroxide, adjusted to pH 2; temperature, 70 °C; UV detection at 217 nm. CZE: UV detection at 214 nm; buffer, acetonitrile-10nM disodium tetraborate (15:85), adjusted to pH 11.7 with sodium hydroxide; capillary, 90 cm × 50 μm I. D.; applied voltage, 25 kV (i = 14 μA).

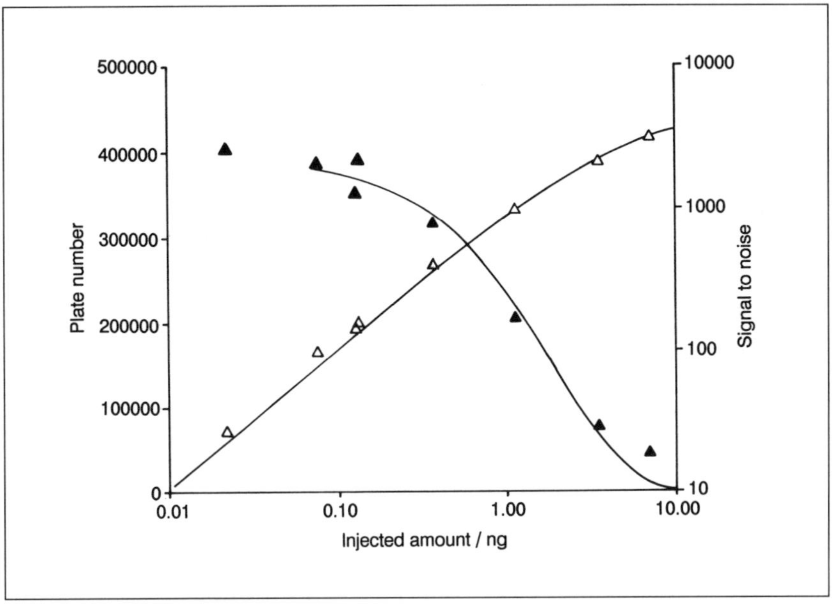

Fig. 3: Effect of amount injected on efficiency and signal-to-noise ratio in the CZE of terbinafine with constant injection volume.
▽ = Number of theoretical plates; △ = signal-to-noise ratio.

2.3 Method development time

The time required to develop a separation method depends not only on the number of parameters which are available for the modification of the selectivity [6], but mainly on the time required for column equilibration, the performance (N/t) and the efficiency (N). CZE shows the most promise in this area because of its very high performance and efficiency (Fig. 1) compared with SFC and HPLC. SFC offers a higher performance than HPLC and often shorter equilibration times [9, 10], because following changes of the supercritical fluid density a stable system is rapidly obtained. From the above it can be concluded that HPLC requires the longest method development time.

However due to the high degree of automation over night optimization providing a global optimum is possible in HPLC, also allowing the large number of mobile phases with a wide range of polarity and pH, to be exploited.

2.4 Sample preparation

Ideally, the sample preparation should be carried out independent of the following separation step. The separation should be insensitive both to the

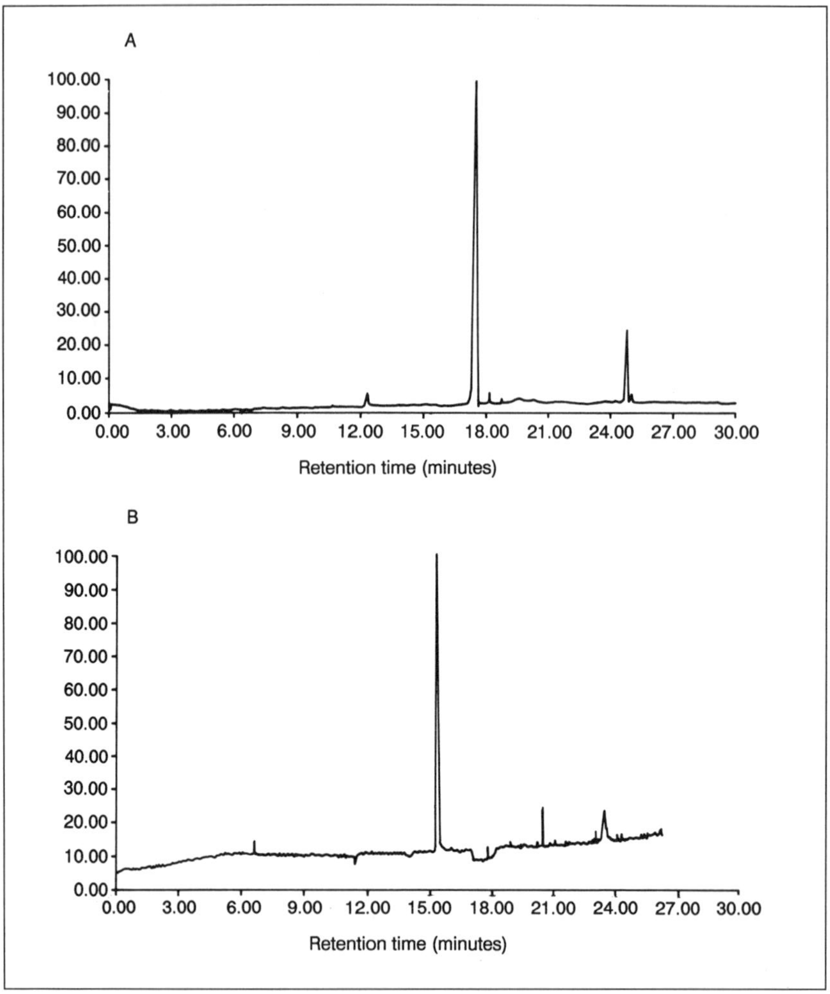

Fig. 4: Interaction between the preparation of the sample and the subsequent separation step. Extraction from dosage forms: (A) CZE, buffer of pH 11.7; (B) standard HPLC procedure. Extraction buffer for HPLC, acetonitrile-water (50:50); buffer for CZE, see Fig. 2. HPLC buffer causes disturbances of the electric field.

nature of the solvent and to the presence of matrix components. When comparing HPLC, SFC and CZE, only Reversed-Phase HPLC comes close to fulfilling these requirements. In this instance only minimum constraints are imposed on the extraction solvent and it may even contain large amounts of water. The choice of extraction solvent in SFC is, in contrast, severely restricted. Polar modifiers containing water would cause spurious artefacts or at the very least a loss of efficiency. Therefore, SFC is often not suitable when polar solvents must be used for the extraction of polar substances from an excipient matrix, or from biological matrices.

In CZE, the sample solvent and the electrophoresis buffer are preferably the same, otherwise disturbances of the electric field or precipitation of the extracted components could occur. The problem can be avoided by preventing electroosmosis and using electromigration as the injection principle, although care must be taken to ensure that the analytes are ionized in the sample solvent. When the separation requires buffers that are vastly different from the extraction solvents, such as extreme pH values, or pure water, the extraction may be disturbed. This is demonstrated for the extraction of a drug and its degradation products from a capsule mixture (Fig. 4). At the necessary pH of 12 the excipients were partly dissolved as colloids, which hindered the centrifugation. In most instances these problems are less critical in CZE than in SFC.

2.5 Technical feasibility

In HPLC with packed columns injection systems are reliable with a reproducibility of 0.5% (relative standard deviation). There are no severe problems with the delivery of eluent in conventional HPLC. Flow-rate relative standard deviations are less than 0.5%. Problem with solvent delivery are only encountered with micro-HPLC systems (column diameter < 1 mm). Fast HPLC with submicronparticles (< 2 μm) is limited owing to the maximum pressure possible of 400 bar.

The injection precision of SFC is relatively poor in comparison with HPLC. Relative standard deviations, on injected amount, of up to 5% were observed, mainly as the result of non-constant splitting ratios or sample precipiation caused by density changes occurring during the injection [11]. The use of an internal standard is therefore recommended. The maintenance of a constant density independent of the flow-rate requires precise restrictors and a sophisticated pressure-flow regulating system. No such systems are commercially available for packed or capillary SFC.

In CZE, concentration-dependent detectors, such as UV absorbance of fluorescence types, are 10–20 times less sensitive than their counterparts in HPLC or SFC and the coupling to mass-sensitive detectors is difficult. Depending on the injection method (electromigration, gravitational or pneumatic), relative standard deviations on peak areas of up to 3–4% are typical [12], and bias effects are unavoidable with electrokinetic injection [13]. For this reason in CZE, as in SFC, the use of an internal standard is advisable. If the capillaries are washed with sodium hydroxide between runs, the fluctuations in electroosmotic flow are less than 0.5%.

As shown in Table 2, for technical evaluation HPLC is the most reliable technique, followed by CZE, for which fully automated systems are now on the market. To a large extent the detector still represents an unsolved problem in CZE.

Table 2: Technical Feasibility (Scale from 0 to 10, where 10 = most reliable)

Operation	HPLC	SFC	CZE
Detection	8	8	1
Injection	10	3	6
Delivery of eluent	8	5	10
Pressure limitation	6	2	10
Total	32	18	27

2.6 Orthogonality of information

The objective of combined analytical separations is to obtain non-redundant information from independent systems [14–16]. In the ideal case different systems can be coupled together to give an on-line multi-dimensional set-up [17].

"Retention parameters" were introduced for the comparison of HPLC with CZE and SFC [5]. Figure 5 demonstrates the orthogonality of HPLC and CZE and HPLC and SFC.

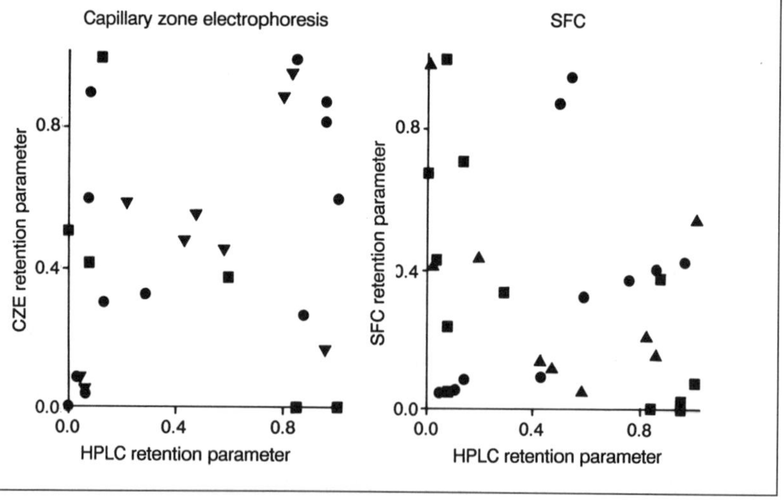

Fig. 5: Demonstration of the orthogonality between HPLC and CZE and between HPLC and SFC. No obvious correlation arises between the retention parameters in HPLC and CZE; some correlation between those of HPLC and SFC may arise (isradipine). All HPLC separations were carried out under reversed-phase conditions, with the exception of isradipine (normal-phase). Compounds: (left)
▼ = terbinafine; ■ = spirapril; ● = AH 21132; (right)
▲ = terbinafine; ● = isradipine; ■ = AH 21132.

2.7 Influence of substance properties on choice of technique

An obvious prerequisite for the analysis of a substance by SFC is an adequate solubility in the supercritical medium. This can often be increased by the addition of a polar organic modifier. However, this leads to a considerable deterioration of the favourable kinetic properties of the supercritical carbon dioxide when the modifier content exceeds ca. 20%. For this reason, substances such as 1.4 dihydropyridins (calcium antagonists), or non-polar excipients (waxes, polymers) which are readily soluble in acetonitrile, chloroform and similar aprotic solvents (comparable with supercritical carbon dioxide in terms of eluotropic strength), are easily separated, whereby only small amounts of polar modifiers (ca. 5%), e.g., methanol, are called for. Moderately basic ionic substances such as alkaloids [18], α-amino alcohols [9], can also be separated, although in many instances suitable ion-pairing agents are required.

CZE in its pure form is suitable only for substances that can be ionized in solution, and therefore requires the use of polar eluents of an aqueous nature, although large amounts of organic modifiers can be added if required. The separation of neutral substances can be achieved using micellar solutions provided that they have an adequate solubility in the buffer mixture.

CZE can be applied both for the separation of low-molecular-weight drugs and also for peptides an proteins.

Owing to the broad scope of possible eluents and stationary phases, HPLC has the fewest constraints. Polar, non-polar, ionic, small and even large molecules, e.g., antibodies, can be separated. Detection problems arise for molecules without chromophores, because of the lack of a simple mass-sensitive detector.

2.8 Optimum use of HPLC, SFC and CZE

HPLC can be utilized in the determination of drugs in biological matrices, where in addition to a high performance a high sensitivity is required. It can be implemented for drug stability investigations in galenical forms, where a reasonably high performance and reliable and fully automated systems are required, for the determination of drug purity and in the analysis of biotechnological products where the use of gradient elution gives rise to a high peak capacity.

CZE is favoured for drug purity control and also for the separation of biotechnological products. It is particularly suitable for the separation of protein digests owing to its enormous peak capacity. Of the three methods, SFC possesses the narrowest range of application in pharmaceutical analysis. Because mass sensitive detection [flame ionization detection (FID)] can be employed, capillary SFC is suitable for the analysis of unpolar excipients.

With a more limited scope than HPLC or CZE, SFC can be used for the purity control of drugs because of its high performance.

2.9 Strengths and weaknesses of HPLC, SFC and CZE

The strength of HPLC is its high technical level, selectivity and versatility according to substance groups and task. The lack of performance and efficiency is partly compensated for by the possibility of using gradient elution. A weak point is the lack of a simple, universal, sensitive method of detection equivalent to FID in gas chromatography and SFC.

CZE displays an enormous efficiency and therefore separation methods can be rapidly developed. The main drawbacks are the poor sensitivity and the fact that elution and extraction buffer have to be similar.

SFC exhibits a higher performance than HPLC and often shorter method development times. An advantage is the possibility of using both modifier and density gradients or a combination of both. Owing to technical problems, SFC with packed columns is far from being at a routine working level. The method is limited to only a small range of eluents and accordingly to a small number of substance groups.

3. Conclusions

From the comparison of the three techniques in drug analysis according to the evaluation parameters discussed above, it can be concluded that HPLC can be used successfully in nearly all areas of pharmaceutical analysis.

CZE has a promising future in drug analysis, where a high number of plates is required within a short time. The method is especially suitable for ionic species. For most applications the detection sensitivity must be improved.

SFC is useful for non-polar and moderately polar compounds or substances for which mass-sensitive detection is required. It is a suitable method for purity control and for the analysis of excipients, although the technical level requires further improvement.

Owing to their orthogonality, CZE and SFC are worth developing, not in competition or as an alternative to HPLC, but as an additional method in order to augment the information obtained from the analysis.

In the future, each of the three techniques will have its place in pharmaceutical analysis, its particular share of application depending on the product range concerned.

Glossary of Symbols

ΔP	Pressure drop		
h	Reduced plate height	A, B, C	Knox equation coefficient
v	Reduced velocity	N	Number of theoretical plates

References

[1] W. Steuer, M. Schindler, G. Schill and F. Erni, J. Chromatogr. **447**, 287 (1988)
[2] J. H. Knox, Chromatographia, **26**, 329 (1988)
[3] J. M. Davis and J. C. Giddings, Anal. Chem. **55**, 418 (1983)
[4] M. Martin, D. P. Herman and G. Guiochon, Anal. Chem. **58**, 2200 (1986)
[5] F. Erni, W. Steuer and H. Bosshardt, Chromatographia, **24**, 201 (1987)
[6] W. Steuer, I. Grant, F. Erni, J. Chromatogr. **507**, 125 (1990)
[7] W. Steuer, I. Grant, Nachrichten aus Chemie, Technik und Laboratorium **38**, 561–706 (1990) Nr. 5
[8] J. P. Chervet, R. E. J. van Soest, M. Ursem, J. Chromatogr. **543**, 439 (1991)
[9] W. Steuer, J. Baumann and F. Erni, J. Chromatogr. **500**, 469 (1990)
[10] W. Steuer, M. Schindler and F. Erni, J. Chromatogr. **454**, 253 (1988)
[11] M. L. Lee, B. Xu, E. C. Huang, N. M. Djordjevic, H. C. K. Hang and K. E. Markides, J. Microcolumn, Sep., **1** (1989) 7.
[12] D. J. Rose and J. W. Jorgenson, Anal. Chem. **60**, 642 (1988)
[13] X. Huang, M. J. Gordon and R. N. Zare, Anal. Chem. **60**, 375 (1988)
[14] J. F. K. Huber, E. Kenndler and G. Reich, J. Chromatogr. **172**, 15 (1979)
[15] H. Vink, J. Chromatogr. **69**, 237 (1972)
[16] D. L. Massart and R. Smits, Anal. Chem. **46**, 283 (1974)
[17] F. Erni and R. S. Frei, J. Chromatogr. **149**, 561 (1978)
[18] J. L. Janicot, M. Caude and R. Rosset, J. Chromatogr. **437**, 351 (1988)

6 Influence of packaging material on the stability of drug products

K. Krummen, CH-Basel

1. Phases of pharmaceutical development

The development of a new drug product can be divided into several phases that are relevant for the pharmaceutical development as shown in Fig. 1. [1] It is in general in the formulation development phase where packaging aspects begin to play an important role.

2. Stability testing in the formulation development phase

The main objective in this phase is to provide sufficiently stable and adequately packaged clinical supplies so that the quality of the drug product is assured for the duration of the clinical study. The aspects to consider here are: Which packaging type and material to choose, possible compatibility problems. In general the packaging for pharmaceutical products must meet the criteria listed in Fig. 2. They can be summarized as good protection, low interaction and high and consistent quality.

The stability tests in the preformulation phase usually include stress tests and tests under conditions compatible with storage statements on the clinical supplies. Stability testing can be restricted to the one or few packaging types actually used in the clinical trials and for bulk storage.

Simple and efficient stability testing protocols can be set up in order to provide the necessary data for an IND.

- ○ Preformulation
- ○ Formulation development
  IND
- ○ Proposed Product
- ○ New Product
  NDA
- ○ Established Product
- ○ Revised Product

Fig. 1: Phases of product development

● **Low permeation**	Water vapour, Oxygen Loss of solvents
● **Low light penetration**	UV
○ **Low adsorption** (absorption)	Drugs, Preservatives Flavours
○ **Low migration** (leaching)	from plastics, rubber glass
○ **Minimal container attack**	
○ **High quality packaging material** (Low failure rate)	Good sealings No cracks, holes

Fig. 2: Improvement of Stability by packaging material

3. Stability tests in the Proposed Product and New Product phases

When developing new drug products for worldwide marketing, pharmaceutical companies generally strive to get the same product formula registered in all countries, although this aim is not always reached. With respect to packaging such a worldwide uniformity is much less frequent, especially for common products such as tablets and capsules. The reasons for this lie in the differing dispensing habits, preferences for certain packaging types in different countries and local availability of packaging materials and packaging machinery. Ideally and most efficiently, stability testing in the various packaging types could be carried out simultaneously for all the packaging types based on a testing protocol sufficient for registration in all 3 regions, EC, Japan and the US.

Some problems encountered hereby shall be demonstrated on a very simple example: a solid dosage form in a humidity tight packaging.

4. Humidity protection: A stability testing protocol for a solid dosage form for the registration in the EC, in Japan and in the US

Let us assume that for a solid dosage form it was decided based on stability data obtained in the preformulation phase, that a humidity tight packaging is necessary and that the packaging types to be used in the EC-member states, in the US and in Japan have been defined: Either a glass bottle with tight closure, a PVC blister tightly sealed in a aluminium bag or a full aluminium blister. The stability testing protocol should be geared at complying with the stability guidelines issued by the EC, the FDA and the MHW in Japan. [2, 3, 4] The issues, which are relevant in this context, are summarized in Fig. 3.

Batches	**EC** **3**	**J** **3**	**US** **3*)**
Long term storage			
○ Test cond.			
Guideline	25 °C (20–30 °C)	25°	acc. to label
Typical	25°/50–60% 30°/40° 30°/70°	25°/75%	30°+ lower (5, 15 or 25°)
○ Schedule		3, 6, 9, 12 18, 24, 30, 36	3, 6, 9, 12 18, 24, 36
Stress test			
○ Test cond.	Temp. Hum. Light	Temp. Hum. Light	Stress Testing frequently used
Typical	3 temp. (+ low temp.) 40°/75% (cycling)	60° 90% r.h.	40°, 50° 37–40°/75%
Accelerated test			
○ Test cond.	–	40°/75%	–
○ Schedule	–	2, 4, 6	–
* Bracketing allowed for dry products			

Fig. 3: Stability Guidelines for Drug Products

This leads to two "room temperature" conditions, i. e. 25 °C and 30 °C, (plus 5 °C or 15 °C as reserve) and two stress conditions (Fig. 4).

Assuming that 3 dose strengths exist of which 3 batches each are tested, then the total number of samples to be analysed is 3 doses × 3 batches × 3 packs × 22 test points = 594.

Batches	3
Long term storage	
– 2 conditions	25 °C 30 °C
– 8 Time points	3, 6, 9, 12, 18, 24, 30, 36
Stress/Accelerated	
– 2 conditions	40 °C/75% 60 °C
– 3 Time points	2, 4, 6
Sampling Points: 2 × 8 + 2 × 3 = 22 per batch	

Fig. 4: Full testing plan

5. Possible ways to reduce the amount of testing

Nobody would accept such a large and extended testing plan without having challenged its efficiency. Possibilities for reducing the amount of testing are:

a) Use bracketing: highest and lowest dose only,
b) One packaging for each country: Test only one pack per country,
c) Use fractional factorial or similar reduced sampling plan.

There are, of course, many other possibilities and every company will try to find its best trade off. But let us confine ourselves here to these 3 options which suffice to make the point.

Option a) brings about a labor reduction of one third but carries a question mark regarding approvability by the reviewing chemist (in the US, preapproval of the testing protocol by the FDA would be necessary).

Option b) (see Fig. 5) is what is usually done because each packaging type can then be tested according to the specifics of the guidelines of the particular country. This is especially important for the Japanese stability guidelines, where also some particular requirements regarding analytics must be followed, e.g. it is necessary to utilize HPLC methods with internal standard. The main disadvantage of option b) is that the possibility to switch from one packaging to another in any of the countries must be sacrificed and that the reduction of labor is only about 18%.

Option c) leads almost inevitably to testing protocols which are outside the guidelines valid in Japan and the US. The protocol in Fig. 6 is only one of several possibilities. It is presented here in order to stimulate the ongoing discussion about international harmonisation. The protocol features some interesting and attractive properties. The labor reduction is more than 50% but nothing had to be sacrificed: The variation between the batches is tested on every dose level, stress tests are included to detect any differences between the packaging types and to reveal any interactions between container and contents. Long term "room temperature" data are provided for all packaging types.

Country	Long term	Stress/ Acc.	Sampling points	
			per batch	for 3 doses 3 batches each
EC	25°+30°	40°/75% + 50°	20*)	180
US	same	same	20*)	180
J	25°	40°/75% + 60°	14	126
				486

*) No testing after 30 months at 25 °C + 30 °C

Fig. 5: Testing plan: "One packaging per country"

Pack	Doses	Batches	Long term 25+30 °C Sampl.	Stress/Acc. 40°/75%+50 °C Sampl.	Sampling points per batch	Total
Glass	3	3	2×8	2×3	22	198
Alu/Alu	3	1	2×3	2×3	12	36
Blister in bag	3	1	2×3	2×3	12	36
Time points:	Full schedule: 3, 6, 9, 12, 18, 24, 30, 36			(8 Points)		270
	Reduced schedule: 12, 24, 36			(3 Points)		

Fig. 6: Example of reduced testing plan

6. What is necessary that reduced (factorial) testing protocols become acceptable?

Firstly, it is obvious that the acceptance of testing protocols which deviate from the guideline path should be increased and especially, that the 3- or 6 monthly testing schedule in the Japanese and the US guidelines is not taken as a "law". In short: Increase flexibility. Secondly, foreign data should be accepted and assessed based on their scientific value even if the testing protocol deviates from the guidelines in the particular country.

Thirdly, supportive data showing that the manufacturer is able to produce a certain type of packaging with sufficient quality should be accepted and evaluated together with the stability data. For our example this means comparative data about the tightness of the 3 types of packaging. Long-term permeation tests carried out at 30 °/75% using desiccant showed that all 3 packaging types provide practically complete protection from humidity. The daily water uptake rates were 100 times or more below the USP specifications for unit dose containers of class A, which is the class with the most stringent requirements.

7. Effects of water on the stability of solid dosage forms

Water influences the stability of solid dosage forms in various ways and is the cause of many unwanted reactions. Examples are given in Fig. 7. It is important to recognize that water does not only act as a chemical agent, e. g. in hydrolytic reactions, but can provide conditions which favour many other reactions leading to degradation and spoilage.

Fig. 8, 9 and 10 show examples for protein aggregation [5], cis/trans-epimerisation of a ring structure and oxydation (formation of N-oxide of a tertiary amine). The influence of water on these reactions is obvious.

- ○ Hydrolysis
- ○ Oxidation
- ○ Racemisation/Epimerisation
- ○ Aggregation of Proteins
- ○ Change of crystallinity / morphic form
- ○ Change of DR
- ○ Growth of microbes or fungi
- ○ ......
- ○ ......
- ○ ......

Fig. 7: Influence of water on solid dosage forms

	Isomer content (%)	
Initial value	0.90	
after 1 year at	25°/50%	30°/65%
PVC/PVDC	1.07	2.25
Glass bottle	0.95	1.07

Fig. 8: Influence of moisture on Isomerisation of a compound with ring structure

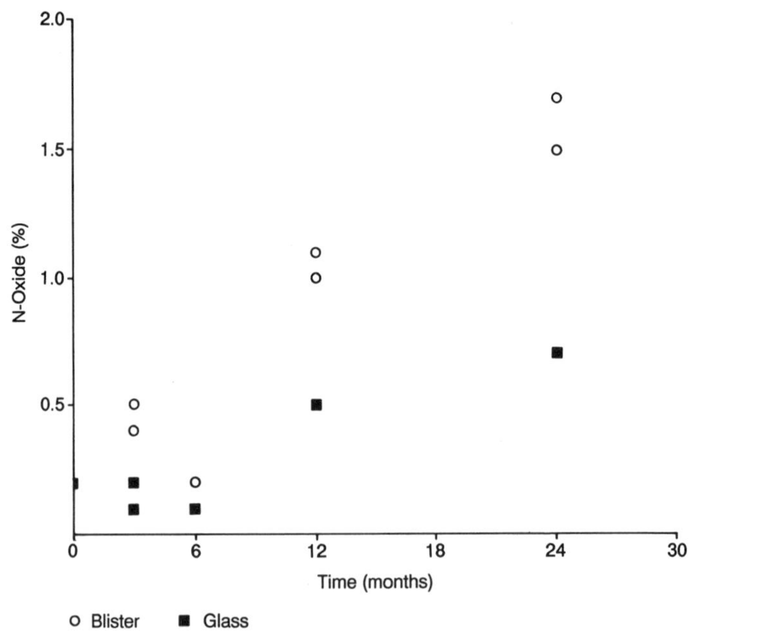

Fig. 9: Formation of N-Oxide in tablets

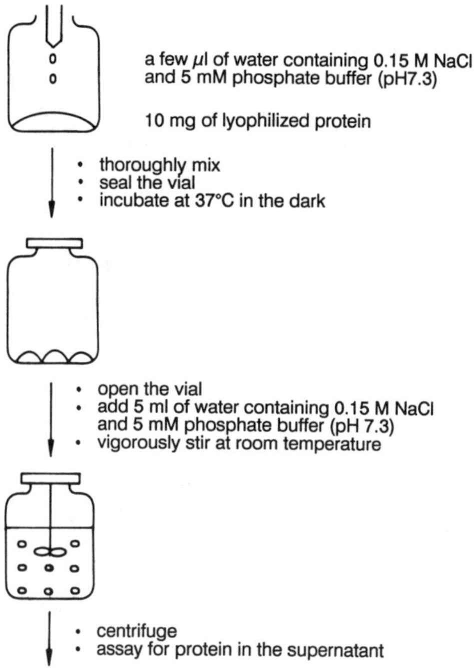

Schematic description of the experimental design developed to investigate moisture-induced aggregation of lyophilized protein in the solid state. See Experimental Procedures for details.

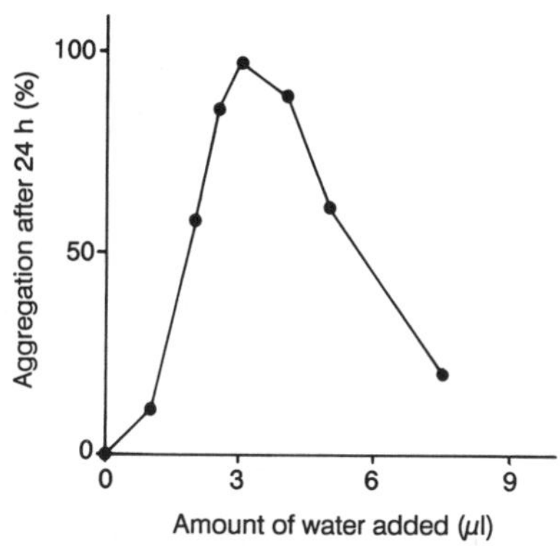

The dependence of the extent of aggregation of bovine serum albumin on the amount of the standard aqueous buffer added followed by 24 h incubation in the solid state at 37°C

Fig. 10: Moisture-induced Aggregation of lyophilized Proteins in the solid state (W. R. Liu, R. Langer and A. M. Klibanov)

The moisture penetration rate is therefore a very important criterion when the packaging material for a sensitive product is chosen.

Blisters made of a humidity permeable, deep drawn film tightly sealed with an aluminium backing, a very popular type of packaging in Europe, are a good example for demonstrating the relation between the humidity penetration rate through the packaging and the shelf life of the product.

8. Blister packagings: Humidity penetration and shelf life

The humidity protection provided by a blister packaging depends on the type and thickness of the blister film material used.

The water vapour permeability of blister film materials is usually expressed in grams of water per m2 per day at a certain temperature and humidity. Typical values are shown in Fig. 11. [6] Such data are very useful for comparing different film materials but it is less obvious how to use them for predicting shelf lives of products. Some basic kinetic principles can help. The moisture transport through a thin packaging film can be described by a pseudo steady-state equation based on Fick's and Henry's law; see Fig. 12. The absorption isotherm of tablets and capsules is approximately linear within a restricted range of the relative humidity, see Fig. 13. Combining these two equations and integrating

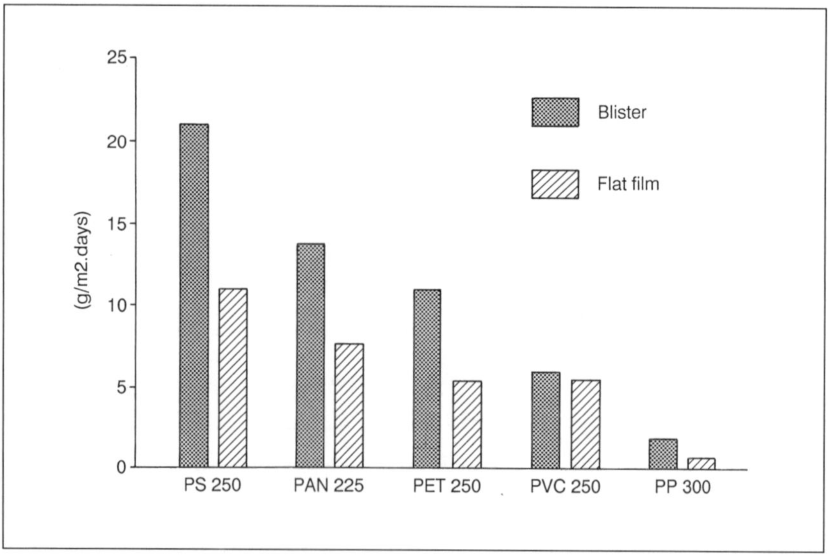

Fig. 11: Mono-Films (WVTR at 38 °C / 90% r.h.)

The moisture transport through a thin packaging film is described by a pseudo steady-state equation based on Fick's and Henry's law:

$$W \, . \, dm/dt = K \, . \, A \, (p_e - p)$$

where dm/dt = rate of moisture per unit initial weight of the medicament transferred per day

K = P / x = film permeability to moisture (mg/day.m2.hPa = mPerm)

P = Permeation constant

x = film thickness

A = Effective area of diffusion (m2)

W = Initial weight of the medicament in one blister pocket (mg)

p, p_e = water vapour partial pressure in the blister pocket and in the environment, respectively

Fig. 12: Moisture transport through packaging film

leads to an exponential function, see Fig. 14. What is most interesting for shelf life considerations is the time period tc.

As a typical example, let us assume a small tablet (100 mg, blister pocket size 1 cm^2), which takes up 1 mg of water when stored unpackaged. The tablet is packaged under dry conditions, e.g. 40% r.h., because the active ingredient is prone to hydrolysis. The initial relative humidity in the pack is then 40%. The degradation reaction may remain very slow as long as the relative humidity in the pack is not higher than 60%. The duration of the protection provided by the

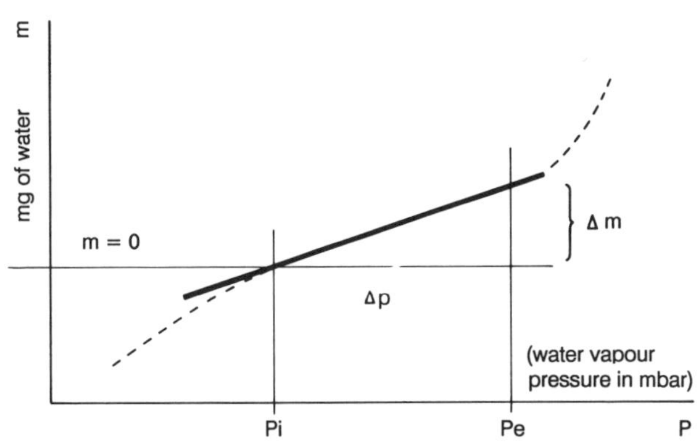

$$m = \frac{\Delta m}{\Delta p} (P\text{-}Pi)$$

Fig. 13: Absorption isotherm

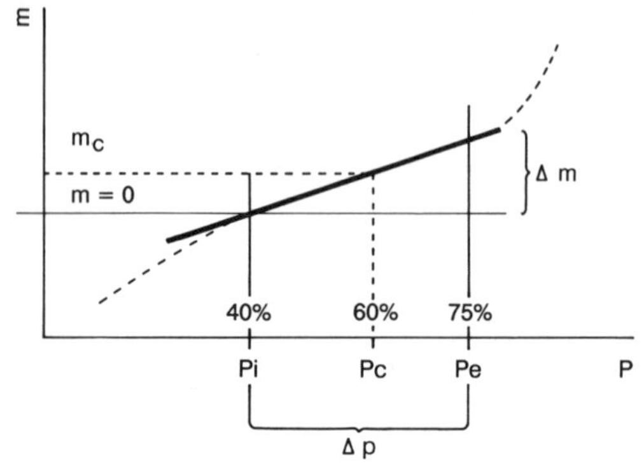

$$m_C = \Delta m \cdot \left(1 - \exp\left\{ - \frac{K \cdot A \cdot \Delta p}{W \cdot \Delta m} t_C \right\}\right)$$

Fig. 14: Critical mass of water

push through pack is obviously the time tc necessary for the relative humidity inside the blister pocket to rise from 40% to the critical value of 60%.

tc is inversely proportional to the permeability K for given values of the maximum weight increase W.Δm and the initial pressure difference Δp. (Fig. 15). In logarithmic scales one gets a straight line of slope -1. This is shown in Fig. 16 for the 3 storage conditions 20 °C/65% r.h., 30 °C/65% r.h. and 30 °C/75% r.h.

As an example, one can see that for a protection time of 3 years in a tropical zone (30 °C/75% r.h.) the permeability of the blister pockets should be 0.5 mPerm. whereas 2.7 mPerm are necessary to achieve the same period at 20 °C/65%.

$$tc = \frac{W.\Delta m}{K.A.\Delta p} \;.\ln\; \frac{Pe - Pi}{Pe - Pc}$$

or

$$tc = \frac{W.\Delta m}{K.A.\Delta p} \;.\ln\; \frac{e - i}{e - c}$$

where

Pe, Pi, Pc = water vapour partial pressure in the environment (Pe), initially (Pi) and at the critical level (Pc)

e, i, c = Percent relative humidity/100 for Pe, Pi and Pc respectively

Fig. 15: Critical time period tc

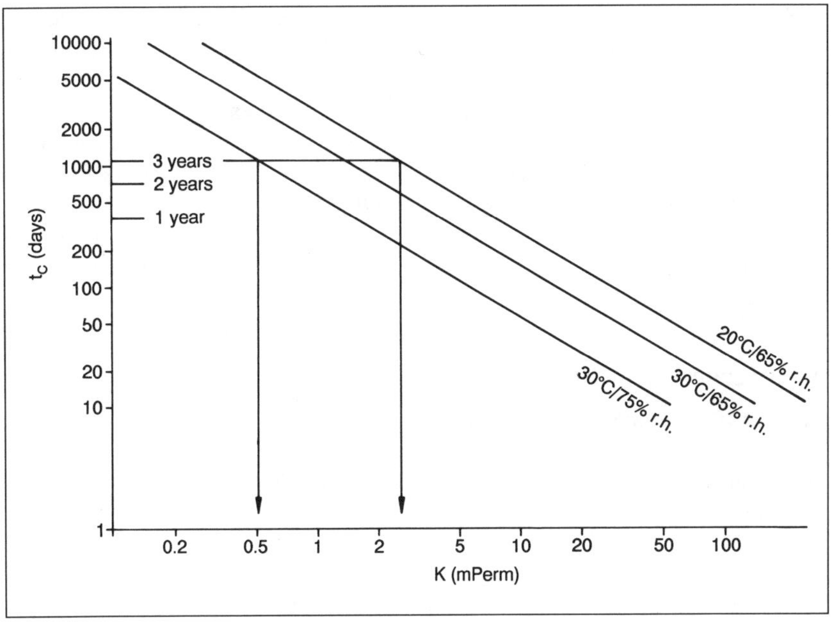

Fig. 16: Influence of Permeability K on the critical time tc

Permeability values of formed films in the range of 2 to 6 mPerm at 20 °C can be obtained with the best film materials currently available which consist of 2 or more (up to 5) layers whereby the barrier layer is made of either PVDC or PCTFE.

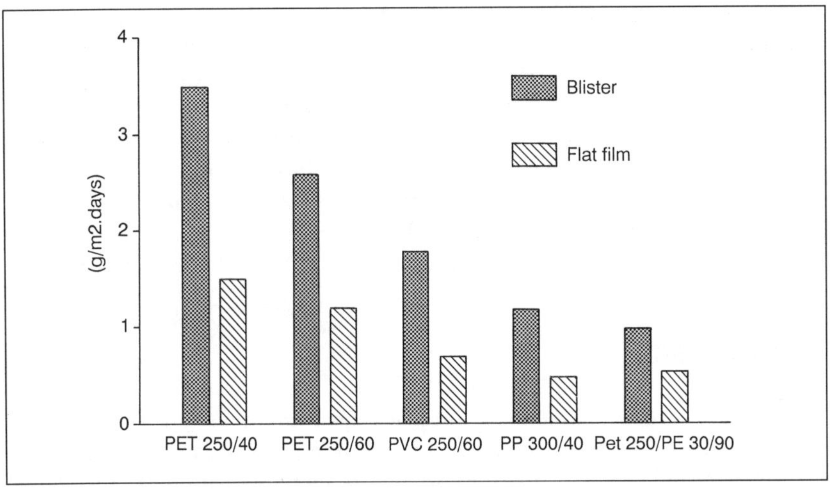

Fig. 17: PVDC-coated films (WVTR at 38 °C / 90% r.H.)

However, halogen containing film materials provoke increasingly criticism in the public, especially in Europe. Fig. 11 and 17 show the water vapour transmission rates of potential alternatives with no or low halogen content. [6] Replacement of PVC as main layer by halogen free films, such us PET, but keeping the 60 μ PVDC barrier layer, provides a reduction of more than 80% of the chlorine compared with PVC/PVDC (DUPLEX) film. The PET/PVDC film can be run on the same blister forming machines as DUPLEX. If one takes PP instead, there is no coating of PVDC necessary in order to achieve the same barrier properties as DUPLEX, but the blister forming machines must be adapted.

9. Interaction between container and packaged pharmaceuticals

Low interaction is a second important requirement for the packaging material. It has two aspects: Adsorption and leaching.

9.1 Adsorption

Packaging materials, especially plastics and rubber, may adsorb or absorb components of the pharmaceutical product. A well known example is the loss of preservatives. Fig. 18 shows the result of a compatibility study between 4 preservatives and 3 rubber stoppers. Stopper C causes high losses of the preservatives already shortly after manufacture of the vials, whereas stopper B is only incompatible with chlorobutanol. Stopper A does not absorb either one of the preservatives. Such compatibility tests must be carried out with the final composition of the pharmaceutical product because the degree of absorption depends usually on the pH and on the other components of the formulation.

The drug substance itself can also be absorbed. Well documented in the literature are the absorption of drug substances from infusion solutions by the plastic bags and tubings used for the infusion. Another important problem is the absorption of flavours by plastic containers. Usually, a change of the odour and or the taste is observed followed by a fading, because the individual components of the flavour are absorbed differently regarding rate and extent.

9.2 Migration/Leaching

The migration of components of the packaging materials into the packaged product has become more and more important in recent years because the use of packaging materials of organic origin has increased dramatically. Plastic materials may contain monomers or oligomers, traces of catalysts, plasticizers,

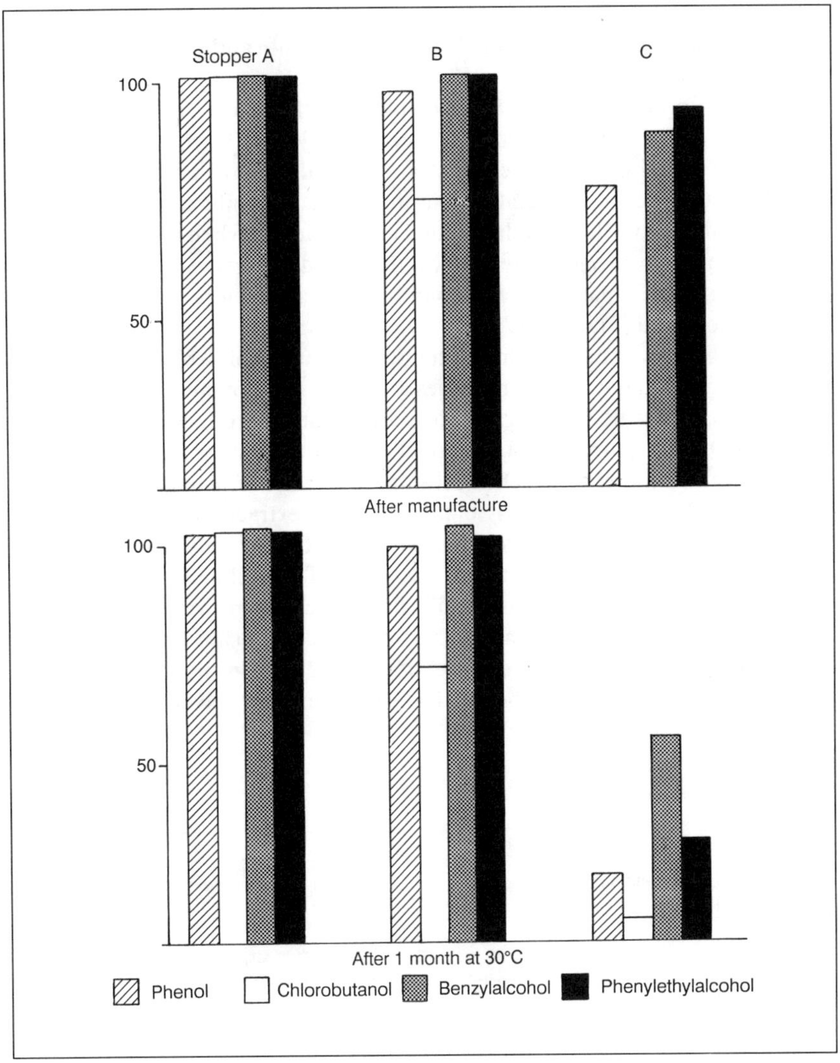

Fig. 18: Adsorption of preservative on rubber stoppers

stabilizers, lubricants, substances added to improve processability and others.

Rubber materials are also potential sources of leachables, e.g. fatty acids, volatile lipophilic substances and heavy metals.

Fig. 19 shows the effect of heavy-metal ions leached from a rubber stopper into a nasal solution containing a drug substance that can degrade by oxidation. Two types of glass bottles and two kinds of rubber stoppers were tested and compared with the same solution filled in ampoules. All solutions were gassed with carbon dioxide. The type of glass had no effect whereas one of the stoppers led to significant degradation and leached also significantly more heavy-metal ions, especially zinc.

A difficult problem observed with lyophilisates for injection is the formation of opalescent or turbid solutions after reconstitution in the solvent. One cause are volatile lipophilic substances which are leached from the rubber stopper and adsorbed at the surface of the lyophilisate. These lipophilic substances remain undissolved or subcolloidally distributed in the solution when the lyophilisate is dissolved and form aggregates which then lead to opalescence. The main component of the volatile leachables from butyl- and chlorobutyl rubbers was recently identified (Fig. 20). [7, 8]

The problem was recognised by the manufacturers of rubber stoppers and new products with improved properties are under development.

Methods for predicting migrants to pharmaceutical products were proposed which are based on exaggerated extraction with various solvents and subsequent identification of the leached substances with modern analytical methods, such as GC/MS, LC/MS, FTIR etc. [9] Soxleth extraction of the polymeric material with solvents of differing polarity and with relatively large amounts of the contacting packaging material was used. The rationale is that such a pronounced exaggeration of the equilibrium partition ensures that any potential migrant is detected. Any migrants detected at significant levels can then be investigated further in the actual situation, i.e. either in the product itself or under the conditions used in the official test methods for solvent extraction.

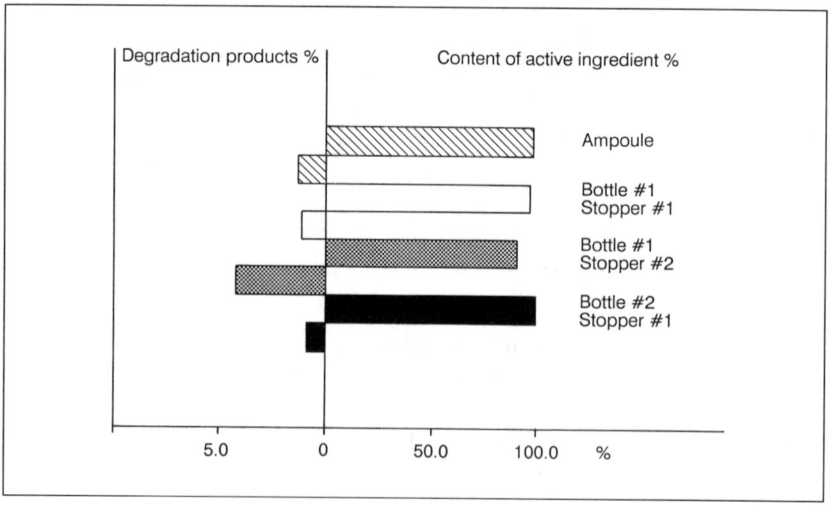

Fig. 19: Nasal Solution (Stressed 5 Months at 50 °C)

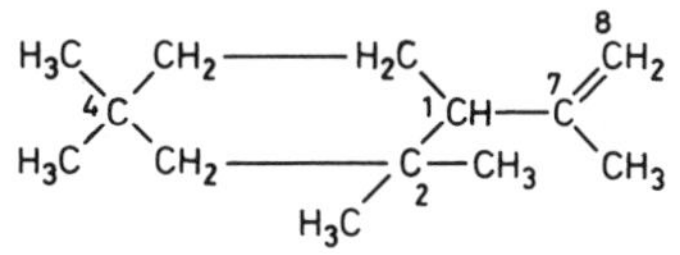

Fig. 20: Main component of volatile leachables from butyl rubbers (R. Jähnke)

References

[1] "Stability Concepts" edited by PMA's joint QC-PDS stability commitee; Pharm. Technol. June, 1984

[2] EC: "Stability tests on active substances and finished products" (Doc: III/66/87-Rev. 4)

[3] J: Stability Guidelines, MHW, 1991

[4] US: Guideline for Submitting Documentation for the Stability of Human Drugs and Biologics, FDA, Febr. 1987

[5] W.R. Liu, R. Langer and A.M. Klibanov, Biotechn. and Bioeng. **37**, 177 (1991)

[6] U. Gfeller, AEL-Facts (Aerni-Leuch AG, Bern, Switzerland), März 1989

[7] R. Jähnke, "Migration von Fremdstoffen aus elastomeren Verschlußelementen für Parenteralia; Diss. Frankfurt a. M. (1988)

[8] R. Jähnke et al., Acta Pharma. Technol. **36**, 139 (1990)

[9] H. Kim-Kang et al., J. Pharm. Sci. **79**, 120 (1990)

7 Plastics for drug products: analysis, possible interactions and official requirements

F. Pellerin, F-Paris

Plastic containers for packaging various pharmaceutical forms of drugs can be used safely only after container-content interactions have been investigated. These migrations are responsible for a wide variety of modifications; release of constituents of the material (low polymers, adjuvants) may modify the stability of the dosage forms or active ingredient, thereby causing intolerance or toxicity. On the other hand, migration and binding to the container of volatile constituents of the finished product may modify stability and preservation of the product and cause degradation of the container material.

Migration phenomena are due to a variety of causes whose identification may to some extent guide investigations. This implies participation of the analytic chemist who uses these data to select the methods most appropriate for achieving detection of migrations. Furthermore, some interactions vary across pharmaceutical forms.

Drug packaging must comply with technological, physicochemical, and regulatory requirements which govern the selection of polymers and materials. Furthermore, technological considerations are essential to ensure the quality and safety of use of these packaging materials.

Two key issues govern the use of plastics for drug products: constancy of the composition of the material and manufactured articles, packaging material and container-content compatibility evaluated by studies of interactions and migrations.

The purpose of this paper is to emphasize these two issues, which require extremely exacting investigations. These, however, should be carried out under realistic conditions.

After a brief review of regulatory requirements in the EEC and Pharmacopeias, which forms the first part of this article, constancy of the material and container-content interactions are discussed in the light of examples concerning semi solid dosage forms and fairly new pharmaceutical forms (patches).

1. Regulations and required data

The European Guidelines and Recommendations to Applicants give a short list of data on packaging material which should be provided in New Drug Application files (table 1); an explanatory note for guidance is being developed and

Table 1: Rules governing medicinal products

Standard format for applications in EEC notice to applicants vol. II (II.C.3. p. 54)
Packaging material (immediate packaging) 1. Specifications and routine tests • type of material • construction • quality specifications (routine tests) and tests procedures 2. Scientific data • development studies on packaging materials • batch analysis, analytical results
Development pharmaceutics and process validation (vol. III p. 9)
Container • appropriate studies should be performed to demonstrate the integrity of the container and closure. A possible interaction between product and container may need to be considered. **Sorption to container** • Data should be presented to show that consideration has been given to the possibility of sorption of the active constituent(s) and additive(s) from liquid or semi-soild formulations if relevant to safety or stability. These phenomena are known to occur with rubber closures and with both glass and plastic containers and administration sets, the data sheet should include an appropriate reference to this fact.

studied. We are participating in the elaboration of this document, we will not repeat the recommendations especially since approval of regulatory authorities has not yet been obtained. We will focus on showing how analytical chemists should approach studies of plastics: data to be collected, selection of methods, and interpretation of results.

1.1 Regulatory requirements in Pharmacopeias: importance of monographs

The European Pharmacopeia is currently an irreplaceable source of data on plastics for pharmaceutical and medicosurgical uses. In 1965, the eighth edition of the French Pharmacopeia emphasized the need for determining the composition of plastics used to make containers, including adjuvants, and for guaranteeing constant quality. For our part [1], we participated in the work which followed this general presentation and led to the publication, in the ninth edition of the French Pharmacopeia, of monographs on high and low density polyethylene, polypropylene, and materials made from plasticized poly (vinyl chloride). This work set the angle of approach and served as a basis for the European Pharmacopeia which discusses a considerably larger number of materials. The list of monographs and articles on plastics in the second edition of the European Pharmacopeia is subtantial (table 2).

The tenth edition of the French Pharmacopeia includes monographs and Propharmacopeia notes which are also valuable sources of information (tables 3 and 4) without commenting on each of the monographs published or being developed, we suggest a classification according to the objectives these texts aim at achieving. Three main groups can be distinguished. The first- and largest-includes the specific monographs. Based on both rigorous experiments and on experience, they describe quality and constancy of quality for each of the plastics or containers they deal with. These monographs consequently form an **irreplaceable collection of data** for the drug industry. In many cases, they leave the user free to choose the additives best suited to each particular case; they require identification of components using Pharmacopeia methods and use of tests designed to guarantee compliance with quality criteria and constancy of composition.

A second, smaller group of monographs, deals with plastic containers and closures. Selection criteria and selection of a type sample to which the selected

Table 2: Monographs in the european pharmacopoea

Materials used in manufacture of containers
• materials based on plasticised PVC for containers for human blood, blood components and aqueous solutions • materials based on plasticised PVC for tubing for blood and blood components
Polyolefines
• Polyethylene las density for containers for parenteral tophttalmic preps • Polyethylene high densitiy for containers for parenteral use • Polyethylene for containers for parenteral use
Silicones
• Silicone oil used as a lubricant • Silicone elastomer for closures and tubing
Containers
• Glass containers for injectable preparations • Plastic containers – Plastics containers and closures – Containers for blood and blood components – Sterile plastic containers for human blood and blood components – Emply sterile containers of plasticised PVC for human blood and blood components – Sterile containers of plasticised PVC for human blood containing an anticoagulant solution – Sets for the transfusion of blood and blood component • Plastic containers for aqueous solutions for intra venous infusion • Rubber closures – Rubber closures for containers for aqueous preparation for parenteral use

Table 3: Monographies francaises publiees ou adoptees pour un prochain supplement

Monographie	Édition
• Colorants des materiaux en matieres plastiques ou en elastomeres a usage pharmaceutique et medico-chirurgical	
	(10ème édition – Juillet 1990)
• Copolymeres (Ethylene/Acide metacrylique)-Zinc	
	(10ème édition – Juillet 1986)
• Feuilles de conditionnment en polyamide	(10ème édition – Janvier 1989)
• Feuilles de conditionnement en poly (Terephtalate D'Ethylene)	(10ème édition – 1990)
• Materiaux en matieres plastiques ou elastomeres a usage pharmaceutique et Medico-chirurgical	(10ème édition – 1990)
• Necessaires pour perfusion	(10ème édition – Janvier 1986)
• Polyamide 6	(10ème édition – Janvier 1986)
• Polyamide 6–6	(10ème édition – Janvier 1986)
Polyamide 6 en fils fins de deametre inferieur ou egal à 1mm	(10ème édition – Janvier 1986)
• Polyamide 6–6 en fils fins de diametre inferieur ou egal à 1mm	(10ème édition – Janvier 1986)
• Polyamide 6–10	(10ème édition – Janvier 1986)
• Polyamide 11 (P) pour feuilles	(10ème édition – Janvier 1986)
• Polyamide 11 (K) pour feuilles	(10ème édition – Janvier 1986)
• Polyamide 11 (P) pour moulage	(10ème édition – Janvier 1986)
• Poly (Ethylene/Acetate de vinyle) pour Recipients destines a contenir des Preparations injectables aqueuses	
	(10ème édition – 1987)
• Poly (Terephtalate D'Ethylene)	(10ème édition – 1987)
• Poly (Terephtalate D'Ethylene) en fils fins de deametre inferieur ou egal A 1mm	
	(10ème édition – Janvier 1986)
• Polytetrafluorethylene pour catheters	(10ème édition – 1989)
• Polytetrafluorethylene opaque aux rayons x pour catheters	(10ème édition – 1990)
• Recipients en matiere plastique destines au conditionnement des collyres aqueux	(10ème édition – 1990)
• Recipients en materiau a base de PVC plastifie pour solutions injectables aqueuses pour perfusion	
• Avis aux fabricants de specialistes pharmaceutiques concernant l'utilisation d'un nouvel additif dans les materiaux pour recipients en matiere plastique a usage pharmaceutique	
• Materiaux en polyethylene	
• Materiaux en polypropylene	

material is compared are defined; the need for a constant composition is emphasized. Another article published as a Propharmacopeia note N° 395 (table 5) lists the data to be provided in applications for use of a new additive in plastics. From a more general viewpoint, we believe the technical data recommended are valuable for pharmaceutical firms studying new materials.

Table 4: Notes techniques pro pharmacopoea en preparation

- Materiaux a base de poly(chlorure de vinyle) non plastifie pour conditionnement de formes seches
- Materiaux a base de PVC plastifie pour conditionnement des solutions aqueuses non injectables
- Feuilles en matiere plastiques simples ou complexes destiness au conditionnement primaire du materiel medico-chirurgical sterile, des articles de pansement, des sutures et de formes pharmaceutiques ou au conditionnement secondaire des recipients souples de solutions steriles

A third, also small group, includes monographs or drafts defining a constituent, for instance coloring agents used in plastics. Indeed, it seems unfeasible to develop a monograph for each for the multiple coloring agents used in plastics. Conversely, a general monograph defining basic qualities promotes safety, while leaving users responsible for selecting appropriate agents and looking for release of this agent in the drug.

1.2 Use of plastics is not confined to the materials or containers described in the Pharmacopeia monographs; on the contrary, alternative compositions given in some specific monographs expand their field of application. Furthermore, the Pharmacopeia offers a particularly broad spectrum of analytic methods capable of readily solving most practical problems. However, the

Table 5: Reproduction de la note technique propharmacopoea n° 395: donnee pour etudier l'introduction d'un additif dans les materiaux en matieres plastiques ou en elastomeres a usage pharmaceutique et medico-chirurgical

1. **Donnees generales**
   - Compose largement utilise dans plusieurs pays, notamment europeens
   - Interet technologique de l'emploi de l'additif; exemples d'applications, de materiaux et de materiels le renfermant

2. **Donnees techniques**
   - Definition, structure, nature, solubilite du compose propose comme additif
   - Methode d'identification precise et specifique applicable dans la monographie concernee
   - Comportement et degradation dans les conditions de fabrication des objets manufactures tels que transformation lors de l'extrusion, stabilite dans le temps
   - Comportement four l'emploi pharmaceutique ou medico-chirurgical par exemple, aux divers modes de sterilisation
   - Donnees generales sur les problemes de migration
   - Donnees toxicologiques a court et long termes
   - Donnees bibliographiques comportant notamment les avis d'organismes nationaux, europeens et internationaux en precisant exactement le conditions generales et les usages particuliers pour lesques l'homologation a ete accordee et plus specialement en ce qui concerne l'utilisation dans les emballages pour produits alimentaires ou il y a un contact direct et prolonge entre l'aliment et le materiau

Pharmacopeia deals only with the plastics most widely used in the pharmaceutical and medicosurgical industries and intended for well-defined or even specific uses (containers for injectable preparations, for blood products, for eye drops; polyethylenes for packaging sheets). It cannot cover the entire range of needs and the pharmaceutical industry also uses other packaging materials. Emphasis should be placed on the fact that – in contrast to beliefs held by some – the Pharmacopeia does not prohibit use of materials or containers with different compositions or with specifications analogous to those of the Pharmacopeia.

For such materials and containers, the pharmaceutical firm develops its own monograph [2, 3, 4] on the basis of physicochemical, toxicologic and biologic (if appropriate) data and defines the selection criteria reported in the Scientific File. In such cases, the Pharmacopeias serve as a general basis for preparing the monograph, and as a guide for developing specifications. Furthermore, Pharmacopeia methods of identification, testing and assay are usually applicable to specific cases. Whether or not, the material or container is included in the Pharmacopeia, the pharmaceutical firm is responsible for undertaking compatibility and stability studies for plastic materials and drugs, and a general study should be carried out.

2. Studies of the materials: constancy of composition

A constant composition is the first requirement for plastic materials and containers. This implies development of analysis protocols and use of validated methods for checking that each bach meets selected specifications. To ensure contancy of composition, a scientific and practical study is needed to justify the routine control test protocol and form a basis for precision and container-content interaction studies.

2.1 Materials made of polymers have especially complex compositions, and it is enough to recall here a few of the various adjuvants and additives used for synthesis or manufacturing processes: plasticizers (polymers or esters), stabilizers, antioxidants, lubricants, "processing aids"... as well as the additives needed for technical reasons or during use of the product: protective agents, coloring agents, charges, etc... "Positive lists" have been developed for food products (EEC guideline 85/372) and are extremly helpful [5, 6] as concerns physicochemical and toxicologic criteria; these lists help select the compounds best suited to the manufacturing of materials and articles and promote safety. However, they cannot be used in the same way as in the food industry [7]. Indeed, for food products, all the compounds on the list can be used, whereas for pharmaceutical purposes these products are not interchangeable. The composition must be constant. Any change in composition requires a new container-content interaction study and in some instances a new stability study.

2.2 From a regulatory viewpoint, implications of the above are as follows:

- Manufacturers of plastic materials should either use only those additives approved for a given product or gather physicochemical and toxicologic data justifying the use of the selected additive and describing its behavior. Data acquired at the European level for food products, although valuable, serve only to predict simulated migrations.
- Manufacturers should inform users of the additives present in the product.
- Manufacturers cannot modify the formulation without notifying it to the pharmaceutical industry.

2.3 In practice, the analytical chemist plays a central role at every level. In this, as in all the other fields of chemical activity, an analyst participates in the monitoring of each stage and indicates the methods best suited from conception to use. At the synthesis stage, analysis focuses mainly on verifications or identifications of structures and on detection of traces (catalysers, low polymers); during transformation, analytic investigations mainly study technological characteristics. At the user stage, analysis centers on constancy of composition, reactivity, purity criteria, and detection of migrations. Analytic methods thus provide an accumulation of data which serve to define quality and routine controls carried out by the user.

Knowledge of all the stages preceding manufacturing of the drug is not necessary. Only those data with an incidence on quality and safety are required.

Participation of all at each level is indispensable. Each participant indicates the characteristics that the final pharmaceutical user needs to know. All collected data describe the material with precision and are included in the scientific development. The responsibility of each participant is therefore involved at every level.

Lastly, the file on physicochemical scientific data needed to select a material must include the following sections.

2.4 General format of a monograph

- Definition of the polymer and nature and role of the main additives.
- General characteristics and solubilities.
- Identification, in particular by infrared spectrophotometry.
- Tests:

- residual monomer and low polymers
- identification of additives (antioxidants, including plasticizers) present in the material, with determination of limits
- traces of metals: Pb, Cd, Zn, heavy metals . . . and catalyser residues.

For pharmaceutical uses, we also recommend a test for resistance to water exposure (25 g in 500 ml boiling water for 5 hours) with determination of UV absorbance, reducing substances, and buffering capacity.

This type of test, included in Pharmacopeia monographs, is useful for evaluating the quality of materials and comparing materials of similar types and compositions.

2.5 Overview of analysis methods

These determinations can easily be repeated on a routine control basis. During product development, complementary determinations can be considered.

Infrared spectrophotometry [8, 9] is the method of choice for identifying a material. Advances in apparatus designs, data analysis (in particular using Fourrier transformation) and computer data banking have considerably improved results. Even with a fairly simple apparatus, relative proportions of vinyl acetate and ethylene in various types of materials and containers made of poly (ethylene vinyl acetate) (E. V. A.) can be determined by calculating the ratio of band intensity.

Treatment of a material or container by dissolution-precipitation [10] also provides a rapid solution to the problem of identification; complete dissolution of the polymer and additives in appropriate solvent, such as tetrahydrofuranne (P. V. C.) or chlorated solvents (polyolefines), is followed by precipitation of the polymer by ethanol. Examination of the insoluble phase by infrared spectrophotometry and of the solution by various spectral (IR, UV) or chromatographic techniques, identifies constituents of the material and very often provides quantitative results of adequate precision. In particular, antioxidants can be identified in the solution using HPLC on C18-grafted silica [9, 11] with tetrahydrofuranne/acetonitrile (75/25) as the mobile phase. Thin layer chromatography as described in the European Pharmacopeia is also capable of separating and identifying these constituents. Other procedures can be used and TLC can be adapted to suit the multiple compositions encountered [12, 13, 14, 15, 16].

Gel permeation chromatography (on a PL 10^5 A gel column) after calibration is capable of determining the molar mass of the material and of measuring the polydispersion index which is predictive of the material's sorption properties [16].

The apparatus is composed of a high performance liquid chromatography chain with a specific column whose pore size is similar to the size of the compound analyzed, a detector appropriate for the chemical structure of the material (refractometer or ultraviolet spectrophotometer), and a GPC integrator; this last is first calibrated using standards of known molar masses and is then used to construct the log curve of molar masses according to elution volume and to determine the molar mass and polydispersity index of an unknown compound.

Applications are multiple and involve

- Polymers used in plastics for pharmaceutical or medicosurgical uses and polymers used as vehicles for drugs, such as polyethylene glycols, carbopol, etc...
- By running the product through several columns of different pore sizes, a single injection enables to evaluate not only the material's molar mass, but also the presence of adjuvants (e.g., antioxidants and preservatives) [17].
- Liquid and gas-liquid chromatography.

The use of H.P.L.C. and G.L.C., techniques, permits to identify and quantify the additives, and also to study their migration and interaction, and, in some cases, to measure their stability when subjected to oxydizers or radiations [9].

As example, we proposed a method using a combination of chromatographic methods (H.P.L.C. – G.P.C.) infra-red techniques with Fourier transform and ^{13}C, ^{1}H Nuclear Magnetic Resonance (N M.R.) to bring confirmation of the structures and the composition of Copolymers polymethylmethacrylates (PMMA) – polyvinylchloride (PVC).

Observation of the characteristic bands of the I.R. spectra indicates the presence of both the structural elements of PMMA (1732 cm^{-1}) and PVC (612–696 cm^{-1}) as shown table 6:

By exclusion chromatography, elution peak having an average molecular mass of 40000, is fragmented at the column exit. Closer examination of I.R. spectra allows to conclude to the presence of one copolymer or two polymers having similar molecular mass.

Nuclear magnetic resonance of ^{13}C and ^{1}H put into evidence the CH level corresponding to V.C about 0,1 and 2 p.p.m.; the CH_3 of MMA acid in corresponding esters appear clearly 3,6 p.p.m and 0,7 to 1,2 p.p.m respectively. The examination by ^{13}C.N.M.R. conforms these results.

Table 6: Characteristic bands of the I.R. spectra

1732 cm^{-1}	Aliphatic ester R-COO-R'
1270 cm^{-1} 1240 cm^{-1} 1192 cm^{-1} 1149 cm^{-1} 750 cm^{-1} 826 cm^{-1}	Polymethacrylic ester
612 cm^{-1} 696 cm^{-1}	C-Cl Band

3. Interaction studies

Studies of interactions between plastics and the drugs they contain are not part of the field covered by the Pharmacopeia. However, the Pharmacopeia describes a limit test for migration of di (2-ethylhexyl) phtalate (DOP) in the monograph on plasticized PVC bags containing an anticoagulant and intended for collecting blood.

European guidelines are not extremely stringent since they call for an interaction study if needed and of studies of sorption and leaching.

In fact, interaction studies, i.e. studies of container-content compatibility, need to be adapted to each drug formulation. Consequently, we present only a few general recommendations based on our experience [23]. We will successively discuss technologic criteria for the article and pharmaceutical forms.

3.1 Technologic criteria for the article

The quality of a package depends on a number of technological criteria which are defined and verified at the stage of transformation into a container or manufactured article, as well as at the stage of pharmaceutical development.

3.1.1 Factors influencing migration

Migration phenomena are related to a variety of causes [18, 19, 20] including thickness of the plastic, contact surface area, and permeability to gases and vapors; the sorption phenomenon is one of the most important of these factors. Absorption occurs mainly in amorphous areas and facilitates penetration within the material of compounds absorbed on the surface; adsorption mainly affects crystalline areas which are sites of optimal absorption and adsorption until a steady state characterized by saturation of all the binding sites is reached; furthermore, dipole-dipole bonds, hydrogen bonds and Van der Waals forces cause interactions between macromolecules and chemical compounds. Formation of these bonds depends on parameters inherent to the composition of the products, including nature of solvents, concentration of the main constituents, pH of the medium, and pK of acid-base couples (acidity promotes sorption).

Interactions are due mainly to the dissolving properties of constituents and additives; formation of covalent bonds and reciprocal solubility are also responsible for some migrations [18].

Thus, a host of factors are involved.

- Crystalline and chemical structure of the material
- Physical characteristics of the plastic.

The analytical chemist must, consequently, measure the influence of a number of parameters, including pH, solvents, temperature, concentration and plastic additives.

In addition to these factors, modifications due to changes in technological properties must be taken into account: changes in volume, formation of defects, thickness of the plastic, shaking, filling.

Lastly, other modifications are caused by treatments to which the drug is subjected, especially in case of sterilization, and by duration of storage.

3.1.2 Systematic approach to technological modifications

Verification of the constancy of technological criteria implies acquisition of a substantial amount of data during development of the drug; ideally, this study is carried out on the drug in its final package. Usually, experience acquired with several types of material of similar compositions guides the final choice.

A few accelerated tests can be carried out under reduced pressure (200 mmHg) at 45 °C.

Effects of the main treatments are evaluated by qualitative observations: collapse, lengthening, swelling, yellowing, exudation of product, cracking.

These observations assess variations in shape and weight which reflect permeability and its variations.

3.1.3 An impressive list of verifications to be performed during selection and development studies is thus obtained. For instance, permeability of a plastic over time may vary because of gradual accumulation of moisture in the package and, after some time, of the slow but inevitable loss of plasticizer; this phenomenon may increase during storage of drug packaging materials.

Similarly, under some conditions, molded plastic articles may crack or split with wear.

These cracks are due to slow changes in composition or structure of the plastic under the influence of chemical, thermal or mechanical constraints.

There is much room for thought about packaging of products in temperated climates, such as that found in France, with temperatures below 0 °C during the winter. Plastic articles subjected to mechanical stresses become deformed regardless of the material they are made of. Similarly, fluage is an irreversible deformation which persists after removal of a mechanical stress applied to a plastic article for a sufficiently long period of time.

Fluage is, in fact, increased deformation due to viscous flow causes by stresses.

Fatigue should be taken into account for plastic articles subjected to alternating and repeated stresses.

The fatigue limit can be determined on the basis of the material used and of the shape of the article; it is defined as the maximum, general, alternating stress this article can withstand indefinitely at constant temperature without breaking. Ultimately, all these physical parameters influence the life span of the material and should be taken into account.

3.2 Chemical approach to migrations

3.2.1 Chemical approach to migrations

Interactions between plastics and drugs can be studied under various conditions even if only the analytical problem of migration is taken into account. The study of interactions is based on previous data acquired during development of the product and, concomitantly, during stability tests.

Because of the large number of pharmaceutical formulations, a global study of migrations or simulations are not sufficient and would only give an unwarranted impression of safety [7].

The European Guidelines (85/572/EEC) suggest four simulation fluids for plastics intended for contact with food products:

A. Distilled water or water of equivalent quality
B. 3% (m/v) acetic acid in an aqueous solution
C. 15% (v/v) ethanol
D. Rectified olive oil or, as appropriate, sunflower seed oil or synthetic triglycerides; the recommendations define the criteria which should be met by each of these products.

For pharmaceutical uses, these studies have only a predictive value and should be completed by a specific container-content compatibility study.

Migrations related to contact with plastic materials, articles or containers give rise to a wide variety of reactions which affect both the strength and quality of the material and the properties of the drug. Container-content incompatibility, release of constituents of the material, and intolerance are problems which can be solved only on a case-by-case basis. A few physicochemical data can serve as a starting point for the development of a study protocol intended for checking compatibility.

In addition, many factors must be taken into account including aggressiveness of excipients, long periods of contact, and storage conditions; changes in technological criteria and physicochemical compatibility must be taken into account, and methods appropriate for each case must be used. These tests must be carried out at each step of drug packaging and storage and should address the effects of sterilization if appropriate.

Migrations are related both to solubility and to the diffusion coefficient of constituents of the material dispersed throughout the polymer and of constituents of the liquid final product; determination of the apparent separation coefficient provides information on solubilities and serves as a working hypothesis (518).

Interactions are studied by successive approaches based, in particular, on the incidence of reciprocal solubilities of the constituents of the material, and of the various excipients. It is often convenient to study the excipients in the absence of active ingredients, which may interfere with the various chromatography and spectral studies. Nevertheless, the final study should be carried out on the finished product.

Three examples of pharmaceutical forms, where migration problems are especially significant, are discussed: paste-like forms, illustrating the role of excipients and flavors in migrations, injectable forms illustrating the role of sterilization, and transdermal therapeutic systems (TTS) where migration of some of the constituents of the system may occur.

3.2.2 Semi solids dosage forms

Migrations from container to content are common with semi solid dosage forms, i. e. ointments creams for topical use and suspensions, syrups...

A large number of additives and excipients used in the pharmaceutical industry are polymers; reciprocal dissolution of the polymer may occur. Under these conditions, migrations may be significantly increased. Excipients present in fairly large amounts may also be involved because of their ability to dissolve constituents of the material, or because of their affinity for the polymer; this is the case for polyoxiethylene glycols, their eters, polysorbates, and generally speaking additives, thickeners, emulsifiers, etc...; properties can be substantially modified if precise specifications and purity criteria are not developed for these products.

In addition, many natural products (gums, mucilages, starches, modified starches) are macromolecules whose contact with the plastic container may give rise to an interaction. In general, the importance of functional groups (hydroxylated phenolic or alcoholic functions for instance) cannot be overstated; these groups may act because of their affinity for water, dissolving properties, or chemical reactivity.

General study protocol

The large number of compounds needed to prepare the product, their chemical nature and their dissolving properties are all causes of migration which compound the study of interactions.

Interactions between excipients and containers are studied using the complete formula minus the active ingredient, as well as partial formulas which facilitate evaluation of migrations; in this way, the constituents of the excipient responsible for the migrations can be identified and grouped, according to their role in the preparation. Furthermore, this method also facilitates the analytic phase involving detection of traces by sensitive techniques such as UV spectrophotometry and HPLC. An example of fractions studied is complete formula - placebo - fixed excipients + flavors - fixed excipients - preservatives and additives present in low levels.

In every case, concentrations and proportions of the various constituents are identical to those found in the finished product.

Excipients or fractions are placed in plastic containers, and stored at room temperature, and at 45 °C for 3 to 6 months.

Furthermore, 2 × 2 cm fragments of the material are placed in flasks made of neutral glass containing the solution and stored under the same conditions as above. The material/formulation ratio is increased approximately twofold in this test.

Analysis method

Analysis of contents consists in detection of any antioxidants released and, in the case of polyterephtalates, of terephtalic acid or low polymers.

The formula under study is extracted by hexane.

The hexane phase is examined by UV spectrophotometry between 230 and 300 nm.

Antioxidants can be detected by liquid chromatography on a column of C18-grafted silica (7 u) with tetrahydrofuranne/acetonitrile: 25/75 as the mobile phase and a UV detector.

Plastics, exposed to the product under study and containing the additives and migrating products, are extracted with methanol.

The solutions are examined by UV spectrophotometry between 230 and 300 nm or by HPLC.

Additives in pharmaceutical forms are detected using HPLC and identified on a C18 column using as the mobile phase methanol/acetonitrile/acetic acid/water (10/10/10/70) for sorbic acid and methanol/water for hydroxy-4 benzoic esters.

Migration of flavors in the container is detected by gas-liquid chromatography on a glass column 3 m long and 3.2 mm in internal diameter, filled with 20 M Carbowax on SWHP 80/160 mesh Chromosorb; detection is achieved by flame ionization.

Results [5, 11]

- Migrations of constituents of materials in preparations.

For polyolefines, generally speaking, tests after storage at room temperature, and at 45 °C for 1 to 3 months have failed to detect migration of constituents of the materials under study; however, for polypropylene containers filled with syrups or emulsions, slight migration has been evidenced with an absorption band corresponding to BHT and a concentration of 100 ppm in the pharmaceutical preparation.

Migration of some compounds towards the plastic is seen under a variety of conditions. Examples include sorbic acid, parahydroxyenzoic esters, and a large number of constituents of topical creams and ointments [18].

- Migrations of flavors in materials.

Migrations of flavors into materials is a well-documented phenomenon; in general, some of the constituents bind to the material and the most volatile constituents are eliminated, causing modifications in organoleptic characteristics to occur.

The various constituents of the flavor have different affinities for the material; for instance, during storage the ratio of menthol to menthone contents of mint essence becomes reversed.

It should also be pointed out that some topical creams are packed in aluminium tubes whose inner surface is coated with resins such as epoxyphenolic, vinylic, and polyamide resins.

Tests for migrations under normal and aggressive conditions are especially necessary for liquid preparations containing organic solvents.

3.2.3 Parenteral forms

- Migration in the injectable solution.

Migration of plasticizers, of plasticized PVC or of additives, especially antioxidants, has been addressed by a large number of studies. For plasticized PVC materials and containers used to hold blood and blood products, the European Pharmacopeia recommends detection of and sets a limit for di (2-ethylhexyl) phtalate. Because this compound cannot be looked for in blood, the Pharmacopeia has developed a migration test in a solution of dilute ethanol whose ability to extract phtalic esters has been shown to be identical with that of blood.

Protocols for interaction studies have been proposed for injectable pharmaceutical preparations. Organic solvents and, in general, excipients and adjuvants used in parenteral preparations have marked dissolving properties with respect to plastics and can quickly cause traces of additives to appear; the main compounds released are traces of phtalates for plasticized PVC materials used for tubings and phenolic antioxidants for polyolefine syringes.

Oily solutions should not be allowed to contact plastics made of polyethylene (because of solubilities) or of plasticized PVC (extraction of phtalic esters), even for very short periods of time. For injectable preparations containing up to 3% benzylic alcohol, this compound migrates into the syringe material and vice versa, and increases migration of the antioxidants contained in the plastic towards the injectable preparation. Migration of constituents of elastomeres, such as vulcanizing agents, should be studied extensively since non-negligible traces may appear after short contact times with the injectable solution.

Lastly, organostannic stabilizers should not be used in containers made of PVC with migrating plasticizers, such as phtalic esters; they are soluble in the plasticizer and, consequently, also migrate into the injectable solution.

Analysis method

For infusion lines, the analysis protocol involves circulating the injectable solution previously diluted in saline in the tubing to be studied, under the conditions of an infusion (15 ml). The solution is collected at regular intervals in a fraction collector at the end of the line and measures are carried out on each fraction.

For syringes, the injectable preparation is placed in contact with the syringe and, for multipart syringes, with each part (barrel, piston, elastomer piston tip). Experiments are performed over 2 hours, 8 hours, and 25 hours, at room temperature and at 45 °C. These parameters are selected to reproduce conditions in continuous infusion pumps and to create aggressive contact in order to facilitate detection of migration towards the material.

For syringes and infusion lines, the experiments are carried out with the following solutions [5]:

- Complete injectable preparation used mainly to study migration of the content towards the container.
- Solvent and additives in the injectable preparation for studies of migration of the content towards the container, phtalates (plasticized PVC), antioxidants (polyolefines), various additives used in elastomers.

- Migration of active ingredients to platics

Migration of active ingredients to plastics is a problem mainly for infusion lines made of plasticized PVC. With some drugs, such as trinitrin (trinitroglycerine), significant binding to materials occurs.

Results show that 90% of the expected amount of trinitrin which should have been perfused during the first hour is fixed on the plastic material.

For some drugs, binding is absent, or minimal, and occurs only during the first hours, and without doubt trinitin is an extreme case, but binding to infusion lines is not an uncommon occurence with other drugs, such as benzodiazepines for instance. Generally speaking, an equilibrium between two phases can be hypothesized a circulating, fluid phase and an immobile phase made of plasticized PVC with a phtalic ester.

Analysis method

The study protocol involves circulating the injectable solution previously diluted in saline, or isotonic glucose, though a PVC infusion line under the normal conditions of an infusion (15 ml/min). The solution is collected, at regular intervals, in fraction collectors at the end of the line, and the active ingredient is assayed in each fraction using HPLC or UV spectrophotometry.

Incidence of sterilization

The study of container-content migrations with drugs includes determination of the behavior of containers during sterilization. The Pharmacopeia rightly recommends that tests be performed on objects previously subjected to sterilization [14, 21].

A large number of plastics cannot withstand heat sterilization. Beta or gamma rays cause structural modifications of polymers, such as cross-linking, and consequently modify mechanical, technological or physicochemical properties. Radiosterilization also affects the additives present in plastics.

Thus, phenolic antioxidants sterilized by beta or gamma radiation are degraded. Degradation products of BHT and irganox 3114 have been detected by HPLC and the corresponding chemical structures determined by mass spectrometry [22, 23, 24]. Forced irradiation of irganox 3114 with 10 Mrad causes formation of an oxidation derivative. The structure of this compound, determined by mass spectrometry using electron impact and chemical ionization, corresponds after scission of the molecule to di-tert-butyl-3,5 hydroxy-4 benzaldehyde. Similarly, degradation products obtained by subjecting samples of BHT to 2.5, 5, and 10 Mrad have been examined by HPLC and mass spectrometry [11, 18, 24, 25, 26].

The content of these degradation products is correlated with the radiation dose. Structures, as determined by mass spectrometry, correspond to the BHT dimer and trimer. Confirmation of this structure, in particular of the presence of a hydroxyl group, can be obtained by chemical derivation using phenyl isocyanate and HPLC analysis.

Ethylene oxide sterilization should take into account this product's reactivity with water, halogens, and functional groups of organic molecules. The problem of elimination of ethylene oxide traces has now been solved by degassing under stringent conditions of temperature and duration, but polymers and their additives undergo multiple transformations: changes in the polymer's chemical structure (polymerization, color...), oxidative degradation of antioxidants, modifications in technological criteria (hardness, plasticity...), or physical criteria (alteration of the surface and of absorbing properties...).

3.3 Transdermal therapeutic systems

Control tests for detection of migrations with transdermal therapeutic systems (TTS) raise analytic problems.

Generally speaking, these systems are composed of several plastic parts which are in contact with the drug (active substance and excipient) and with the skin.

Obviously, only plastics with well-defined compositions can be used in TTSs. Analysis is fairly simple since the different parts can be studied separately, taking into account the part they play in the system: barrier effect (isolation from the environment) or permeability effect (penetration of the active substance).

3.3.1 In addition to determination of characteristics, identification tests and assays, complementary studies of mean molar mass and polydispersion of molar masses are often needed to characterize polymers; size exclusion chromatography is appropriate for these studies.

For copolymers, such as poly (ethylene-vinyl acetate) (EVA), the relative proportions of monomers should be defined. For some materials made of plasticized PVC, the di (2-ethylhexyl) content should be determined. Additives, plasticizers, antioxidants, and stabilizers, even present in small amounts, should be identified and assayed, since they may migrate in the substance and be found in the drug along with the active ingredient and excipient.

Excipients are often macromolecules also [28]. In this case, mean molar mass and polydispersion should be determined as for other polymers. Furthermore, purity criteria should be strictly defined; in addition to the impurities listed in the Pharmacopeia, these compounds often contain small amounts of antioxidant preservatives which may migrate in the system or even pass through the skin. It should be pointed out that the "appropriate for foodstuffs" label is inadequate since the rout of administration is not oral. Development of TTSs should therefore include extensive investigations of the various plastics and excipients used.

Safety can be guaranteed only on the basis of compatibility studies. Because of the large number of materials used, it is difficult to propose a general study protocol. On the basis of experience, it seems that studies of migrations between each plastic used in the TTS and the excipient, under aggressive conditions, can predict exchanges to some extent; furthermore, prior knowledge of the composition of materials used is obviously helpful for guiding investigations.

The sensitive methods used in analytic chemistry (GLC, HPLC) are capable of detecting traces and lowering detection thresholds. Data acquired then form the basis for deciding whether a toxicologic study and detection of products capable of crossing the skin in body fluids are warranted.

Conclusion

Prior knowledge of the nature of the polymer and of the main additives present in plastic containers, measurements of physical factors influencing interactions and evaluation of their roles, and physicochemical studies of migrations and exchanges form the basis for container-content interaction studies. Each migration is, however, a unique phenomenon, and data obtained with multiple samples are not helpful for solving interaction problems and developing a general rule. Only predictions, based on structures and on physical and chemical properties of drugs and materials, can be made. A general study protocol to be adapted to each individual case and complemented if needed can be proposed.

The examples given show that the analytic approach to plastic-drug interactions is fairly easy; chromatographic and spectral methods are accurate and reliable means for detecting traces of migrations.

Because of the considerable development of platics and innumerable variety of formulations of materials and objects, a specific interaction study is needed for every material that contacts a drug; only in this way can the physicochemical analysis effectively contribute to the development of plastics and offer its own guarantees which, together with toxicologic and clinical studies, ensure safe use of drug packaging materials.

References

[1] Pellerin, F. et Mancheron, D., Ann. Pharm. Fr. **27**, 65 et 469 (1969)

[2] Pellerin, F. article dans Postaire, E., Matières platiaques à usages pharmaceutique et médicochirurgical, Editions Médicales Internationales Paris, Vol. 1 (1986), Vol. 2 (1987)

[3] Colloque Association Conditionnement et Santé, Paris 1986

[4] Colloque INSERM, Ministère de la Santé, Paris 1985

[5] Pellerin, F., Dumitrescu, D., et Andre, D., Sci. Techn. Pharm. **10**, 127 (1981)

[6] Pellerin, F., et Baylocq, D., STP-Pharma **1**, (9) 847–855 (1985)

[7] Pellerin, F., et Baylocq, D., Ann. Falsif. Exp. Chim. **82**, N° 876, p. 139 (1989)

[8] Tamini, O., Thèse Doct. Pharmacie 3è Cycle, Université Paris XI, 1983

[9] Majcherczyk, C. et Pellerin, F., à paraître

[10] Andre, D. – Thèse Doct. es. Sci. Pharm. (Etat) Rouen 1981

[11] Pellerin, F., Baylocq, D., Majcherczyk, C. et Ai-Laham, Ann. Pharm. Fr., **42**, 15–25 (1984)

[12] Pellerin, F. et Dumitrescu, D., Talanta, **243**, p. 27 (1979)

[13] Pellerin, F. et Baylocq, D., Ann Pharm. Fr. **42**, 15–25 (1984)

[14] Majcherczyk, C., Thèse de 3è Cycle Pharmacie, Université Paris XI, 1985

[15] Majcherczyk, C., Polge, P., Baylocq, D. et Pellerin, F., Talanta, **33**, (12), 985–989 (1986); **24**, 27 (1989)

[16] Baylocq, D., Majcherczyk, C. et Pellerin, F., Ann. Pharm. Fr. **47**, 196 (1989); **43**, 329 (1985)

[17] Yagoubi, N., Baillet, A., Pellerin, F. et Baylocq-Ferrier, D. STP-Pharma **5**, (10) 693 (1989)

[18] Cohen-Goldberg, M., Mémoire de D.E.A. de Dermopharmacie, Université de Paris XI, Laboratoire de Chimie Analytique, Prof. F. Pellerin 1983

[19] Asworth, R.W. et Heard, D.D., J. Pharm. Pharmacol. Suppl. 18, 98 (1966)

[20] Deluca, P.P. et Kostenbauder, H.B., J. Am. Pharm. Assoc. (Sci. Ed.), 49, 430 (1960)

[21] Guillot, F., Legendre, B., Baylocq, D. et Pellerin, F., Ann. Pharm. Fr., **41**, 367 (1983)

[22] Schoefter, F. et Giet, Ph., D.E.A. de Pharmacochimie Option "Chimie Analytique", Université Paris XI, 1985, sous la direction de Baylocq, D., Pellerin, F.

[23] Pellerin, F., Majcherczyk, C., Baylocq, D., Talanta **33** (12), 985–989 (1986)

[24] Pellerin, F. Baylocq, D., Majcherczyk, C., J. Pharm. Clin. H.S. 217 (1988)

[25] Yagoubi, N., Baillet, A., Pellerin, F. et Baylocq, D., J. of Chrom. 522, 131 (1990)

[26] Pellerin, F., Baylocq, D. – STP-Pharma 1, 847 (1989)

[27] Pellerin, F. et Coll.-STP-Pharma **4**, 51 (1988); **5**, 693 (1989)

[28] Pellerin, F. – STP-Pharma, 6 HS, 119 (1990)

8 EC Guidelines for stability tests on active substances and finished products

A. C. Cartwright, GB-London

1. Introduction

The European Community's guideline on stability testing of active substances and finished products was adopted by the Committee on Proprietary Medicinal Products (CPMP) in July 1988 and published in Volume III of 'The Rules governing Medicinal Products in the European Community' – 'Guidelines on the quality, safety and efficacy of medicinal products for human use' in March 1989. Its adoption was the fruit of many months of discussion between the national authorities and with the industry to produce requirements that would enable medicinal products to freely circulate within the EC, both within the temperate Northern European countries and the Mediterranean countries.

This harmonised guideline is important for current national marketing authorizations in the EC and for the creation of a single European market (of 320000000 consumers) in pharmaceutical products after 1992. It is envisaged that applications for certain categories of product (biotechnology & high technology products) will be made through a centralised procedure. Thus, such products will be given (after 1992) a European marketing authorization and this means that product labelling (including storage requirements and shelf-life) must be uniform throughout the Community.

In addition to the EC, CPMP guidelines are also accepted by the European Free Trade Area (EFTA) and Nordic countries. Applications for New Active Substance (NAS) products may often be made to countries which participate in the PER (Product Evaluation Report) scheme. This again assumes a common basis of assessment of requirements such as stability.

Since the publication of the guideline there has been an increasing industrial interest in international harmonization of requirements for marketing authorizations. Stability testing is one of the more complex, time-consuming and expensive parts of the pharmaceutical development of a new product. Different national requirements inevitably mean that tests may have to be repeated for different markets. One of the purposes of this seminar is to consider how there might be some harmonization of requirements which would enable testing to be done using a test programme which would enable companies to generate data which could be used for marketing products in all of these territories. As part of this process of harmonization, there has been discussion over the last 2 years with FDA and the Japanese Ministry of Health & (MHW). The first International Conference on Harmonization (ICH) will take place in November 1991 in Brussels and harmonization of stability test requirements will be an important part of the agenda of the quality workshop of the Conference. This seminar will be a useful preliminary to the ICH meeting.

2. CPMP guidelines – the process of adoption

CPMP Guidelines are written by the CPMP and Commission Working Parties – Safety, Quality, Efficacy & Biotechnology. A rapporteur is appointed from the delegates who produces the first (& subsequent) drafts of the text. There may be some preliminary discussion with the industry associations.

The draft text is then considered by the CPMP and sent out for consultation with the national authorities, the European trade associations (EFPIA and AESGP), the US Food & Drug Administration and the Japanese Ministry of Health & Welfare (MHW). Their comments are considered by the Working Party and the text amended as appropriate. The final amended text is then put to the CPMP at one of its plenary meetings for adoption. The text of the guidelines is then sent to the industry for distribution to their member companies.

When a number of guidelines have been adopted they are published by the CPMP in a volume of guidelines.

2.1 The function of CPMP guidelines

The function of guidelines is to serve as a non-mandatory framework. They are a set of guidance notes from the EC authorities to the industry. They set out:

- the main points to be dealt with in the dossier for the marketing authorization (MA) application,
- the minimum testing requirements (e.g. the number of batches of finished product to be tested,
- an indication (in general terms) of the methods, taking care not to go into technical details that are unnecessary to those skilled in the sciences of product, process and analytical development.
- examples of test requirements for specific products (e.g. to avoid frequently recurring problems in MA applications).

3. The EC stability guideline-definitions

The purpose of stability testing is to enable information to be generated which enables well-considered proposals to be made for the shelf-life of the medicinal products and the recommended storage conditions.

The function of stability test is to ascertain how the quality of a medicinal product varies as a function of time and under the influence of a variety of environmental factors.

The quality of a product is shown as:

- its active ingredient content
- its purity (limitation or absence of degradation products of the active)
- its organoleptic properties
- its microbiological properties

The function of the label, package leaflet and Summary of Product Characteristics storage recommendations is to guarantee the maintenance of the product quality in relation to its safety and efficacy when in use and in regard to its acceptability to the patient. The label and leaflet needs to cover all aspects of the storage, distribution and in-use manipulation of the product (its dilution, suitable compatible containers for IV administration etc.).

4. The active ingredients stability

For new Active Substances (NASs) data is invariably generated by experimental studies.

For existing (known) active ingredients, as a rule evidence from the scientific literature is acceptable but comparative accelerated studies may be needed in certain cases such as

- a change in the route of synthesis
- for actives where there is known to be a stability problem (e.g. coated ascorbic acid used in multivitamin products).

Some of the active substance stability testing may be carried out as part of the Development Pharmaceutics (as described in the CPMP guideline on 'Development Pharmaceutics and Process Validation'). These studies are usually designed to show the effects of different solvents, effects of pH, effects of light oxygen etc.) and are intended to elucidate likely routes of degradation, degradation products. These studies will enable stability – indicating analytical methods for the active substance itself and methods for the separation and quantitation of the degradation products to be devised.

The guidelines recommends that a minimum of 2 batches should be evaluated under a variety of temperature and stress conditions. The active ingredient must be assayed using a validated stability-specific method with identification and quantitation of the principal degradation products.

The studies enable the storage conditions (suitable packaging materials for the bulk substance and recommended storage temperature) to be defined. The equivalent of the shelf-life for the active is the **retest date**, i. e. the date at which the conformity of a batch of active ingredient with its specification must be rechecked.

The data from these studies and the Development Pharmaceutics also enables a preliminary choice of formulation and packaging material as well as suggeting suitable analytical methods.

5. Stability of the finished product

5.1 Climatic conditions in the EC and relationship to test conditions for Long Term Real Time Testing to Stress Testing of the Finished Product

The purpose of stability testing is to carry out tests under defined conditions which will simulate the adverse effects of storage under the conditions to which the product might be subject during distribution handling and dispensing.

The climate in the EC can be divided into 2 zones:
Zone I: Temperate climate
Zone II: Mediterranean-like/sub-tropical climate.

The EC countries in Northern Europe (Denmark, Germany, Belgium, Northern France, Luxembourg and Netherlands) all fall into Zone I (temperature zone). Countries in Southern Europe fall into Zone II (Mediterranean zone).

To generate product stability data for products to be distributed throughout the EC (and EFTA countries), the shelf-life and storage conditions must be based on the 'worst case' conditions, i. e. Zone II storage conditions. It is now no longer acceptable to store product in uncontrolled Zone I 'room temperature' conditions and to pretend that this will give information applicable to Southern Europe. Neither is it acceptable to have dual Zone labelling (i. e. labelling with a different shelf-life for Zone I and II countries).

The two fundamental factors in determining product stability during storage are temperature and Relative Humidity. Haynes (1) derived a formula from the

Table 1: Zones & climate.

Climatic Conditions	Zone I	Zone II
Mean temperature (°C[la])	< 20.5 °C	20.5–24 °C
Mean Annual Relative Humidity %	45	60

Table 2: Kinetic mean temperature

	Zone I	Zone II
Kinetic Mean Temperature (°C)	21 °C	26 °C

Arrhenius equation with which be found it possible to calculate the 'virtual temperature' from individual temperatures during the year to which a product might be subjected. The 'virtual temperature' is the isothermal storage temperature to which a product must be exposed to simulate the non-isothermal effects of variations in normal storage due to seasonal and climatic variations in temperature and relative humidity. This concept has been further developed by Dr. W. Grimm (2) as the 'kinetic mean temperature'.

Cartwright (3) has argued the case that using temperature test conditions which simulate Zone II storage will enable data to be provided for all of the major pharmaceutical markèts, the EEC, USA and Japan.

The EC has now chosen 25 °C as the Zone II standard test condition for Real Time storage. For the finished product, at least 6 months data should be supplied in any application for a marketing authorization.

5.2 Stress Testing

This is invaluable for:

- providing complementary data to support a shelf-life request if long-term Real Time data is not available
- showing the effects of adverse storage conditions on the packaging and product to enable any special storage precautions to be derived

The EC guideline suggests a variety of possible test temperatures and conditions, such as:

- 3 higher temperatures
- low temperatures (e.g. − 15 °C, 2−8 °C and freeze-thaw cycling)
- high humidity (e.g. RH not less than 75%)
- high Temperature & RH
- light (either artificial light or natural day-light)

5.3 Batches to be tested

Normally, **at least** 3 batches must be studied except where the active ingredient and the products is known to be chemically stable, no decomposition products are formed and other properties of the product do not change significantly.

5.4 Packaging material

Sufficient data must be provided on the finished product as proposed for marketing. However, supplementary data on batches stored in related packs may be used to augment these data.

5.5 Product characteristics to be evaluated during stability testing

The CPMP guideline indicates that the characteristics studied should be:

- those in the finished product specification likely to be affected by storage
- those not necessarily monitored routinely in the finished product specification but which it is important to check, e.g. dissolution of oral or vaginal tablets.

Product characteristics can be divided into:

- physical (tablet hardness, in vitro dissolution, particle size of suspensions)
- microbiological (e.g. preservative efficacy at the end of the reported storage period)
- chemical (e.g. assay of active ingredient, content of decomposition products, content of other agents such as anti microbial preservatives).

5.6 Validation of Test Methods

All test methods are required to be described in detail in the marketing authorization application. All methods must be fully validated and the data from the validation studies provided in the application (see separate guideline on 'Analytical Validation'). The method for the active ingredient must be stability-specific.

5.7 Presentation of Data

It is convenient for assessors if the data is summarised in tables and graphs. It is particularly useful for the summary to allow comparison of the inter-batch stability.

5.8 Discussion & Analysis of Data

One of the most intractable problems with applications for marketing authorization is in the section dealing with the discussion, analysis and evaluation of the shelf-life, storage precautions and the check (shelf-life) specification.

One particular requirement in the EC is for a Pharmaceutical Expert Report to accompany the application to provide a critical analysis of the data. Some particular points which need to borne in mind for the Expert Report are:

- 'material balance' (the idea of considering the assay for the active ingredient and the levels of the known degradation products together to see if all of the observed degradation is accounted for)
- 'least stable' batch (if batches of the finished product have a significantly different stability, the shelf-life, overage and storage warning should be based on the behaviour of the least stable batch)
- shelf-life after first opening (a separate label/package insert statement of the shelf-life after first opening may be needed in some circumstances such as if the product contains a volatile preservative)
- shelf-life after dilution or reconstitution (details of all diluents which may be used need to be explicitly stated by their official name (e.g. BP monograph title) and data on both physical stability and assay given to support the recommendations).

5.9 Labelling

The objective is to have a single set of label storage recommendations which can be used in the Single Pharmaceutical Market after 1992 and which will allow products to be moved across frontiers. The guideline indicates

- that if there is evidence that defined storage temperatures are needed, these must be stated on the label (in °C), such as 'Store below 25 °C', 'Store in a refrigerator at 2–8 °C' etc.
- if the product is stable at 25 °C and withstands some weeks' exposure (from stress tests) up to 30 °C, the product need bear no special storage instructions

6. Ongoing Stability

If data on production batches are not provided, ongoing studies should be carried out on the first 2 or 3 production batches and the results provided to the authorities.

7. Toxicity of degradation products

Potential toxicity of degradation products should be reviewed and discussed in both the Pharmaceutical Expert Report and the Toxicological Expert Report.

8. Bracketing & Matrix Design

The EC guideline does not refer to these concepts. However, some companies in the EC have used these concepts to reduce the burden of testing. It is likely that in any future revision of the EC guideline, the place of these concepts will be clarified.

9. International Harmonization

Discussions will continue during 1991 in bilateral discussions with the US FDA and the Japanese Ministry of Health & Welfare to try to improve harmonization of stability testing. This subject, as mentioned, will also be a major part of the agenda of the November 1991 International Conference on Harmonization. This APV conference will enable some useful preliminary review of differences in approach to be considered – particularly with reference to temperature test conditions, the possible need for evaluation of the toxicology of degradation products and the use of validated analytical methods.

References

[1] J. D. Haynes, J. Pharm. Sci. **60** 927 (1971)

[2] W. Grimm, Drugs Made in Germany **29** (1), 39 (1986)

[3] A.C. Cartwright, Drug Development Ind. Pharmacy, **15** (10), 1743 (1989)

9 The revised Japanese Guidelines for stability studies

Yasushi Takeda, J-Osaka

Stability study for new drug application to Ministry of Health and Welfare (MHW) had been generally conducted according to the standard testing procedures contained in the Notifications from March, 1980. There had been domestic and foreign requests for elimination of rigidness from the standards, e.g. meaningless replication of identification by color reaction. In 1989, MHW began to review the existed standards in order to remedy the practical shortcomings and to make the stability test more pertinent to quality assurance of drugs on the market, as well as to promote international harmonization of stability study of drug products.

The revised Guidelines were made public in the Notification from Pharmaceutical Affairs Bureau (PAB), MHW entitled "Handling of data on stability testing attached to applications for approval of manufacture and import of drugs", on February 15, 1991. The Notification consists of two parts. Part 1 is "Detailed policy to deal with stability data", which refers to the policy of PAB on handling of stability data in New Drug Application. Part 2 is "Guidelines for stability studies", which refers to testing methods of stability studies.

English translation of "Guidelines for stability studies" is as follows.

1. Preface

The purpose of the stability studies required in applying for approval of drugs is to evaluate the stability of a drug's quality in order to ensure its efficacy and safety and to establish its storage requirements and shelf life. These guidelines provide the standard methods to conduct stability testings. It is not always required to stick to the standard methods provided in the guidelines, when there are rational reasons for altering testing methods, and scientific bases for believing the objective of stability testings is achievable by the altered method[1].

2. Definitions

2.1 Specimen: a sufficient quantity of a bulk drug substance or drug product taken from a designated lot in order to conduct its stability testings.

2.2 Testing sample: a sample taken from each specimen in order to conduct testings at certain repetition times at each sampling point.

2.3 Repetition times: the frequency of repeating the whole testing procedure with each testing sample in order to keep the acceptable precision of testing results.

3. General considerations

3.1 Not only parameters specified in Specification and Testing Methods of a written application for approval but parameters likely to be affected by storage or those useful for the examination of stability shall be tested.

3.2 Sampling and data analysis shall be made by statistically appropriate methods. Specimens shall be chosen so that they may be representative of all production lots, and testing samples shall be taken randomly from the specimens. The number of testing samples[2] and repetition times[3] shall be determined so that the reliability of testing results may be improved.

3.3 It is not required to stick to the assay method for an active ingredient and its degradation products specified in Specification and Testing Methods of a written application for approval but to choose a stability-indicating method.

3.4 In stability studies of a drug product, it is required to examine whether the potency of additives such as preservatives is maintained.

3.5 When changes in quality are expected due to interactions between drugs and containers, deterioration after unsealing, extreme fluctuations of tempera-

[1] The main idea underlying these guidelines is that if the stability of a drug under the standard storage conditions is ascertained, its stability under milder storage conditions than the standard one will be ensured. In applying for approval, however, it is desired not to stick to ascertaining the stability of a drug under the standard storage conditions provided in the guidelines, but to submit the data which the establishment of packaging or storage requirements is based on. Therefore, it will be one approach to prove the stability of a drug scientifically by utilizing not only data showing no changes in its quality but also data showing changes in its quality.

Exemplification in the footnotes of the guidelines is given to help to explain the objective of testing. The range or value in the footnotes may be modified rationally.

[2] When the degradation rate etc. varies largely among drug products of the same lot but of different packaging units, it is desired to increase the number of testing samples.

[3] In principle, a well-validated assay method producing as little variance as possible shall be employed. When the assay method is not validated very well or when the variance in method is presumably large, it is desired to increase the repetition times.

ture, etc., it is required to conduct stability testings under the storage conditions selected according to the characteristics of a drug product to clarify their effects, as well as under the standard storage condition stipulated for each test.

4. Stress Testing

4.1 Objective

A stress testing is conducted in order to examine the changes in quality under severe conditions that may be encountered during distribution.

4.2 Testing Methods

Specimens and testing samples: specimens and testing samples shall be taken properly from bulk drug substances and drug products. The test shall be conducted with unpacked specimens in principle, but it shall be conducted with packed ones when necessary[4].
Storage conditions:
Bulk drug substance; storage conditions for a bulk drug substance shall be determined so that the effects of temperature[5], humidity[6] and light[7] may be clarified.

Drug product; Storage conditions for a drug product shall be determined so that the effects of temperature[5], humidity[8] and light[7] may be clarified. When the quality of the drug product[9] is presumably affected by extreme changes of temperature and humidity, it is desired to use conditions[10] capable of detecting

[4] When changes in quality of an unpacked bulk drug substance or drug product are detected, the results shall be reflected in Storage or Precautions for Use. When its stability is improved by packaging, it is required to conduct a stress testing with a packed one and ascertain the effects of packaging.

[5] e.g., 60 °C when room temperature is specified in Storage, or 30 °C when a cold or frozen storage is specified in Storage.

[6] e.g., 90% RH. (When loss of water (e.g., from hydrate) is expected to affect its quality, it is required to conduct the testing under low-humidity conditions.)

[7] e.g., 1.2 million lux · hours.

[8] e.g., 90% RH. (When loss of water is expected to affect the quality of a drug product, it is required to conduct the testing under low-humidity conditions.)

[9] e.g., suspensions, emulsions, and semi-solid drug products, etc.

[10] e.g., when room temperature is specified in Storage, it is desired to use conditions where temperature is expected to fluctuate between 5 °C and 60 °C, and those where humidity is expected to fluctuate between 25% RH and 90% RH at 25 °C. When a cold or frozen storage is specified in Storage, it is desired to use conditions where temperature is expected to fluctuate between freezing temperature and 25 °C.

their effects. When a drug product produces different degradation products from its bulk drug substance, their chemical structure, degradation mechanisms, safety etc. shall be discussed.
Storage duration and sampling time:
Storage duration and sampling time shall be determined properly according to the physicochemical properties of drugs so that the objective of a stress testing may be met.
Testing parameters:
Among testing parameters specified in Specification and Testing Methods of a written application for approval, those likely to be affected by storage shall be tested. Further parameters which are essential for examination of degradation products and for quality control[11] shall be tested. (Testing parameters are exemplified in the Appendix.)

5. Long-Term Storage Testing

5.1 Objective

A long-term storage testing is conducted in order to ascertain the stability of a drug during a designated period of distribution.

5.2 Testing Methods

Specimens:
One specimen each of a bulk drug substance or drug product shall be taken from 3 lots. The testing shall be conducted with packed specimens in principle.
Storage conditions:
The specimens shall be stored under the following condition 1) in principle; however, when storage temperature is specified, they shall be stored under the following condition 2)[12]. (When a special storage condition, other than temperature, such as storage in light-resistant containers is specified in Storage of a written application for approval, the testing shall be conducted under the said storage condition, and it is desired to submit the data validating the choice of the storage condition.)

[11] It is required to examine a variety of parameters considered to be necessary for quality control in order to validate testing parameters for a long-term storage testing or an accelerated testing.

[12] When the stability of a bulk drug substance or drug product is likely to be affected by humidity, it is desired to conduct the testing at a fixed humidity. In this case, the storage conditions shall be determined considering the conditions that may be encountered during distribution.

1) 25 °C (± 2 °C)
2) designed storage temperature (± 1 °C)

Storage duration:
Not less than 3 years. (When a expiration dating period is specified in a written application for approval, storage duration shall not be shorter than the period.)

Sampling time:
Sampling shall be made frequently enough to clarify the time course of change in quality. It is usually required to conduct the testing at intervals of not more than 3 months for the first 1 year, and at intervals of not more than 6 months thereafter[13].

Testing parameters:
Among testing parameters specified in Specification and Testing Methods of a written application for approval, those likely to be affected by storage shall be tested. Further parameters useful for evaluating stability shall be tested. (Testing parameters are exemplified in the Appendix.)

Testing sample:
Three samples shall be taken from each specimen[14]. The number of testing samples can be reduced when non-quantitative parameters are tested. Repetition times shall be determined properly according to the precision and reproducibility of the measurement.

6. Accelerated Testing

6.1 Objective

An accelerated testing is conducted in order to estimate, from a short-term testing, the stability of a drug during a designated period of distribution. When

[13] If no change in quality is detected during storage, stability will be ascertained. Even if a change in quality is detected, it is possible to ascertain stability if a chronological trend in the change is properly determined. Therefore, it is desired to test samples as often as possible. If it is known from the results of a preliminary testing that a bulk drug substance or drug product is substantially stable, it will be accepted to prolong sampling intervals.

[14] The width of the 95% confidence limit for a curve of chronological changes (e.g., degradation) in quality is the smallest at a point corresponding to the mean storage duration for all testing samples. Therefore, if chronological changes in quality are detected, it is desired to increase the number of testing samples in the latter testing period so that an expiration dating period may be accurately estimated.

an accelerated testing does not answer the objective[15], a long-term storage testing shall be conducted.

6.2 Testing Methods

Specimens:
One specimen each of a bulk drug substance and drug product from 3 lots shall be taken for each storage condition. The testing shall be conducted with packed specimens in principle.

Storage conditions:
The specimens shall be stored under the following condition 1) in principle; however, when a storage temperature is specified, the specimens shall be stored under the following condition 2)[16].

Other storage conditions shall be included in the testing in addition to the following condition 1) or 2), when necessaryy[17] (When special storage requirements other than temperature, such as storage in light-resistant containers, is specified in Storage of a written application for approval, the testing shall be conducted under the special storage condition, and it is desired to submit the data on which the establishment of the storage requirements is based.)

1) 40 °C (±1 °C), 75% RH (±5%)
2) designed storage temperature + 15 °C (±1 °C), 75% RH (±5%)

Storage duration: Not less than 6 months.

Sampling time:
Sampling shall be made at not less than 4 points of time including the point at the start of the testing.

---

[15] When an expiration dating period cannot be estimated by extrapolating stability data obtained at an elevated temperature (e.g., when the expiration dating period depends on physical properties of a drug which cannot be related to temperature), an accelerated testing cannot be validated.

[16] If it is considered that the accelerated testing of a drug (such as aerosols and suppositories) under the storage condition 1) or 2) is improper, it will be accepted to conduct the testing at a temperature other than the standard condition 1) or 2).

In this case, it is required to choose a storage duration rationally and state reasons for the choice.

[17] When any change in quality is not detected under the storage condition 1) or 2), the objective can be achieved by the testing under the single storage condition 1) or 2). When a change in quality is detected under the storage condition 1) or 2), however, it may be possible to estimate stability at 25 °C from the results of additional testings under other storage conditions (e.g., at 35 °C and 75% RH, or at 30 °C and 75% RH), based on kinetical consideration.

Testing parameters:
Among testing parameters specified in Specifications and Testing Methods of a written application for approval, those likely to be affected by storage shall be tested. Further parameters useful for evaluating stability shall be tested. (Testing parameters are exemplified in the Appendix.)

Testing sample:
Three samples shall be taken from each specimen. The number of testing samples can be reduced when nonquantitative parameters are tested. Repetition times shall be determined properly according to the precision and reproducibility of the measurement.

7. Other considerations

As for the following, it is required to choose a scientifically rational testing method and proper sampling time.

7.1 For drug products which are dissolved or suspended before use, it is required to conduct stability testings with their solutions or suspensions. For a drug product which is potentially combined with other drug products in clinical practice, it is required to examine their stability and compatibility.

7.2 For liquid drug products such as emulsions, suspensions and solutions, it is required to conduct stability testings by storing the drug product on their side or inverted, in order to clarify interactions between containers and contents, and contamination by chemicals eluted from containers.

Appendix: Testing Parameters

Among parameters specified in Specifications and Testing Methods of a written application for approval, parameters likely to be affected by storage shall be tested. Further parameters useful for evaluating the stability of drugs, shall be tested.

It is desired to choose parameters which meet the purpose of each testing (a stress testing, a long-term storage testing or an accelerated testing). Representative parameters are exemplified below.

1 As for bulk drug substances: content (strength), quantities of degradation products, appearance, odor, clarity and color of solutions, optical purity, etc.
2 As for drug products: content (strength), quantities of degradation products, appearance, and characteristics of dosage form, etc.
3 Parameters to be tested according to dosage forms

Moisture: tablets, capsules, powders, solid preparations to be dissolved or suspended before use, etc.
Dissolution or release properties: tablets, capsules, suspensions, suppositories, transdermal preparations, etc.
pH: liquid preparations, solid preparations to be dissolved or suspended before use, etc.
Weight change: liquid preparations and semi-solid preparations in plastic containers, etc.
Particle size distribution: suspensions, emulsions, inhalant aerosols, etc.
Viscosity: emulsions etc.
Eluates: large-volume injections etc.
Particulate matter: large-volume injections etc.
Adhesive strength: plasters etc.
Melting point: suppositories etc.

10 Considerations on revision of the FDA Guidelines for submitting documentation on the stability of human drugs and biologics

C. S. Kumkumian, USA-Rockville

1. Introduction

In the EEC, the need for supplementing drug applications after approval differ from that in the USA. In the USA, major changes need to be addressed by supplements to the approved application which normally need prior approval. For example, changes in formulation, packaging, change in product manufacturing site, and change in specifications and regulatory test methods normally need supplements. Many of these require stability testing. It would be helpful to harmonize the requirements for stability testing for these changes even though the need for supplements varies in the EEC countries. In our revision of the stability guidelines, we expect to clarify to a greater degree the needs for stability testing in various post approval changes.

In this regard, we must keep in mind that the FDA guideline covers the IND, the NDA, and the supplements that are required after marketing approval. It also contains a section dealing with Abbreviated New Drug Applications (ANDA).

Even though there are some differences between our countries in the requirements for drug approval, we should be able to discuss these differences and harmonize more closely. Many of these are scientific and technical questions that we hopefully can come to closer agreement. On the other hand some of these are administrative, social, and should we say the "corporate culture" of the various countries. In discussing these differences we can understand the problems better and improve harmonization.

In our striving for greater harmonization, we are hopeful that drug products will be available to greater numbers of people sooner and possibly at a lower cost.

Keeping in mind these possibilities and the status of our current requirements, we are in the midst of revision of our stability guideline in the USA. One of the important things that is being done is greater clarification of the current requirements. We don't want to make the guideline a recipe book but we want to make it easier to understand our requirements and thereby lessening queries in this regard. We are expanding our section on new drugs as well as that concerning generic drugs.

Stability testing should be relevant to the stage of the development of the drug product. In our view, studies conducted during development of a drug product do not necessarily follow a rigid separation into Phases 1, 2, and 3, but should be adequate for the intended duration of the stability tests.

The emphasis in Phase 3 is on final formulations in their probable market packaging and on expiration dating.

As a point of information, we recently have published a proposal for not requiring an expiration date on clinical supplies as long as there are ongoing stability studies to evaluate the quality of the drug product. Up until this time, this was a requirement under our Current Good Manufacturing Practices regulations.

2. Drug Substance

In regard to the stability studies for the drug substance, we are discussing the possibility to having a retest date on the label. It will also help drug product manufacturers to determine better amounts to purchase over a period of time. After the specifications and tests have been established for the drug substance, along with the packaging, studies should be done to determine the conditions of storage that will assure the quality up to the time of use. Since the time of use can vary, information should be available to inform the purchaser when the drug substance should be checked, such as a retest date. This information should be accrued from studies on three batches of drug substance made by the same process at the same facility. In doing these studies, one would be able to better evaluate the limits of degradation products, if any. The amount of degradation products, as well as with the impurities, that will be tolerated depends on the manufacturing process, packaging, amount of consumption, and the toxicity of the substances. As a result of these studies there may need to be a re-evaluation of the packaging to be used.

3. Drug Product

In regard to the drug product, which usually is a complex system that has been formulated after much development work by chemists, pharmacists, physicians, pharmacologists/toxicologists and biopharmaceutists, there is also a need for assuring the quality of the product until consumed by the patient. In the movement toward greater harmonization the question comes up as to harmonization on the specific stability testing conditions such as temperature, humidity and time. As all of you are aware, temperature and humidity surveys have been done to determine the prevailing conditions in the EEC, Japan, and the USA. These have been divided into four climatic zones. From the information presented, it appears that all three areas, EEC, Japan, and the USA, include climatic Zones I and II. In the USA, there is an area that falls into climatic Zone IV. The U.S. Pharmacopoeia defines Controlled Room Temperature as 15 ° to 30 °C. This definition was determined after much discussion at a public conference held by the USP in the USA. When a drug product is labeled for storage in this temperature range, then we need to assure that the product is stable at the extremes of the range.

To assure the quality of the product under these conditions, we need to carry out stability studies to obtain data that will show us that the labeled storage conditions are correct. The data that is needed usually will include more than one set of conditions. Some of the testing is for a shorter period and at least one test is for long term. In order to confirm an expiration dating period, there needs to be testing under at least one set of conditions for at least the full expiration date. The batches need to be production batches and the product need to be packaged in the container for marketing. Usually, we would expect the long term study for a tablet, for example, to be at 30 °C. We have not normally defined a specific humidity at this temperature. In addition to the long term studies on solid dosage forms we normally receive data at temperatures such as 40 °C/75% humidity and higher. In addition, we do get data at other higher temperatures in many cases. The need for testing at lower temperature and other temperature/humidity combinations will depend on the dosage form.

In some cases it may be possible to obtain adequate data on other variations of testing conditions that may justify the label storage condition and the expiration date. One condition that is somewhat akin to the stability of the drug product, which differs somewhat in the EEC and the USA, is that most of the drug products in the EEC are marketed in unit-of-use or unit dose packaging, while in the USA we still have many large containers that are used by pharmacists for dispensing into smaller containers for the patient. So, in this difference in packaging, our harmonization at the current time is not close and may need to be pursued further. We also need to have the storage temperature stated on the label, even if the drug product is stable at temperatures up to 30 °C.

Another consideration that needs to be addressed in our stability guideline is one that has been stimulated by the proposal in the EEC guide of allowing a product release specification of 95% to 105% of labeled potency without justification. In the USA, we have not addressed the setting of release specifications of drug product, except in certain unstable drug products, in the drug application but have left it up to the CGMP'S and our inspectors. It is not clear how this limit would fare with a monograph for a drug product that has limits of 95% to 105%. Also, I don't see how this range would fare when the expiration dating period is based upon stability data from drug product released at close to 100% of labeled potency.

Another area being considered for clarification is that of accurately documenting the intervals at which the testing was performed. As many of you know, there has been a practice in some firms that when samples of solid oral dosage forms cannot be tested at the designated time period, the samples are frozen and tested later. When the data are submitted for the designated time period, it really does not fulfill the intent of the protocol unless data exist to substantiate that freezing does not affect the results. It would be better to not freeze the solid oral dosage form but to leave it at the temperature station and report the value at the time of analysis. Freezing may not necessarily affect chemical degradation but its effect on dissolution may not be so. For this reason we are considering a revision in the guideline to discourage the freezing practice unless it can be justified by data.

Another consideration for revision is for metered-dose inhalation aerosols which are suspensions. In addition to the regular stability testing conditions, a one-time temperature cycling study is recommended for the purpose of evaluating the effects of temperature change on the quality of the suspension in the container/closure system. Such a study may consist of four six-hour cycles per day between subfreezing temperature and 40 °C for six weeks. In addition, containers must be stored upright and inverted to simulate what may happen in storage and transportation.

Questions have been brought up on the "sterility" testing of sterile products and how to deal with it in stability studies. We have stated that sterility may be assessed by various means, including evaluation of the container-closure integrity by appropriate challenge test(s), by testing for preservative antimicrobial preservatives (if present), or by sterility testing.

Appropriate challenge tests include:

a. dye immersion tests. Sealed containers are submerged in dye solution in either a vacuum chamber or autoclave, processed, then visually inspected. The sensitivity of the test must be validated [1].
b. microbial challenge test with viable challenge population at least 10^6 to 10^8/mL is appropriate for various final containers. Briefly, the medium-filled container is completely immersed in a solution containing a high number of viable (motile) bacteria. The containers are then incubated to detect the growth of microbes that have entered the container through a breach of the container/closure integrity. The ability of the medium to support growth of the challenge organism must be demonstrated [2].

These tests should be designed to demonstrate the ability of the container, seals, closures, and other materials to maintain a barrier to microbial ingress throughout shipping, storage, and product handling. Sterile products that are marketed as prefilled syringes should have data to assure that the syringe continues to function properly throughout the expiration dating period with regard to the applied extrusion force.

Stability studies on reconstituted solutions, such as lyophized products and sterile drug substances, need to be done and the appropriate time period and temperature placed on the label.

Another area that probably needs to be discussed further is that of matrix and bracketing protocols for stability testing. We have not gone into any great detail. We do permit studies of this type as long as justification is provided. This type of testing can be very cost-effective and time-saving when you have many strengths, sizes, and different packages.

For degradation products that arise during the expiration dating period, we need to know the amount(s), the structure of the substance(s), and information on the toxicity of the substances(s), if known. As far as the amounts present, we would get a profile by looking at the amounts present in products during the investigational period. Were these present in the animal studies and clinical trials? We would consult with our pharmacologists/toxicologists on toxicity considerations. If there are limits to be set on individual degradation products, then the normal procedure is to produce that degradation product, probably synthetically, to prepare a Reference Standard for use in the test procedure.

We view that accelerated data for establishing an expiration date is supportive but not the primary data. The primary data comes from the drug product packaged in the market container and stored at the designated temperature(s) on the container label. We will review the accelerated data in connection with the primary data and give a modest extension of expiration date if warranted. Our only exception to this is the case of the generic drugs covered by ANDA's which we will now discuss.

For ANDA's, we are considering quite a revision, primarily a clarification of the requirements. We, in the USA, have had a great influx in the number of ANDA's submitted since the Drug Price Competition and Patent Team Restoration Act passed in 1984. This law permitted generic versions to be marketed under an abbreviated application, when the innovator's patent expires. It permitted marketing based upon the submission of adequate data on chemistry, manufacturing, and controls and also upon adequate bioequivalency data. Because these drug products have been marketed for some time, and usually there is data on the drug substance in the scientific literature, the decision was made that in the absence of full-term stability data for a drug product, accelerated stability data would be acceptable for the granting of a tentative expiration dating period.

The recommended accelerated stability test conditions are:

40 °C and 75% relative humidity, if appropriate, for solid oral dosage forms, suspensions, semi-solid and solutions proposed for controlled room temperature storage (15 °–30 °C):

25 °C and ambient humidity for the above dosage forms proposed for refrigerated storage (2 °–8 °C).

The accelerated stability data should be generated at initial, 1-, 2-, and 3-month test periods. The batch(es) used for this study should be representative of commercial production. If any other stability data are available, they should be submitted. If the results are satisfactory, a tentative expiration dating period of up to 24 months at the labeled storage condition may be granted with a commitment to put the first three production batches on stability studies and submit the data in their annual reports. In addition, there is a need for firms to put at least one production batch on stability under all circumstances.

Once the full term stability data are submitted, the firm can extend the expiration dating period when full term data has been accumulated on the approved stability protocol.

If multiple manufacturing sites are to be used then accelerated data, as described, need to be submitted for each site along with a commitment to place the first three production batches under stability testing. Also, when different sources of drug substance are used, the above criteria are to be used.

As we discuss some of our considerations for change as well as the various differences and similarities between the drug application for the USA and the EEC, we must again realize that both groups have a somewhat different approach to the needs for approval of a drug product for marketing. In the USA, we like to see all the pertinent information, including more detailed data than may be needed for the EEC. In the USA, the reviewers are the persons who write the expert report. We do rely in some areas, such as in the clinical area, on advisory committees to provide expert advice to supplement our internal reviewers. We should diligently strive to improve the process and increase harmonization between the groups. This will hopefully expedite the review and shorten the time needed to bring drug products to market.

As we look at the process of development of drugs, we see that stability testing is one of the more important considerations that need to be addressed. We need to know that the drug products that are used in clinical trials are stable in their quality throughout the investigations. We likewise must assure that the drug products being marketed maintain their quality throughout their expiration dating period.

Because of the importance of assuring the quality and consistency of the drug, the drug regulatory agencies have required the expiration dating of drug products. In order to assure this, pharmaceutical firms must do studies on the stability of the drug. These include studies on the drug substance and the drug product. Studies of this nature are very important and do involve much time and can be quite costly.

Since stability testing is needed for all drug products, it would be considered one of the more important aspects of drug development that need to be looked at more closely in our attempt at closer harmonization. Certainly it is an area where a great deal of queries arise in the drug regulatory process. It is also an area where a great deal of time and manpower are utilized, as well as storage facilities. One of the fruits of harmonization could be the minimizing or absence of redundant testing because of differing regulatory requirements.

In addition to our current requirements and their harmonization, we need to discuss new ways of accomplishing our assurance of drug quality from the experience we have learned in the past decades. These may involve different or new statistical approaches.

One of the questions often brought up is the temperatures at which we store our stability samples and also how long, and at what intervals do we do the testing. The question also arises as to what type of tests need to be done and should they be done at each testing interval. What reliance do we put on accelerated data, Arrhenius equation and statistical projections? What studies are needed before approval and what studies can be done after approval is another question. What are the sizes of the batches needed, the equipment used and whether the product needs to be made in the production department? What role can the inspections play vs the role of the reviewers in regard to submission of data in drug applications? What tests are adequate to assure sterility of parenteral products throughout the expiry dates?

References

[1] Levine, C. S., Validation of packaging operations, (1986) p. 545–593. In F. J. Carleton, and J. P. Agalloco, (ed), Validation of Aseptic Pharmaceutical Processes, Marcel Dekker, Inc., New York

[2] Freiben, W. R., R. J. Flock and A. Derisser, J. Pharm. Sci. Technol. **36**, 112–116 (1982)

11 Stability Testing in Relation to Guidelines in the EC, Japan and USA – A Recommendation

H. Möller, D-Frankfurt/Main

1. Introduction

As a result of the international harmonisation of regulatory requirements, especially the comparable requirements for stability tests in the EC, USA and Japan, pharmaceutical companies can plan and implement stability tests whose results can be used for drug registration in every country of the world. In order for a stability study to be accepted worldwide, it has to conform to an international, that is to say, uniform concept of product quality. In other words, the test must comply with international standards for composition, manufacture, and packaging (primary packaging material) as well as with standardised drug-product specifications.

Stability studies during the development of a drug product are adapted systematically to the given stage of development, the basic aim being to accompany each stage of preclinical and clinical testing with a complete set of stability tests.

Coordinated stability tests enable the quality of a drug substance and drug product to be planned at every stage of development; the tests themselves are classified according to their objectives (Tables 1 and 2). Stability test types I to III are basically concerned with the characterisation of the drug substance and the finished product and are intended to be carried out in conjunction with the preclinical and clinical tests. Once the drug product has been registered, stability tests types IV and V are performed; these tests are designed to ensure that the quality aimed for during development (prototype) is maintained during production, i.e. is reproducible.

2. Stability Tests, Type I

When developing a drug product, the test criteria for the active ingredient should include those given in Table 3 which provide information on the physical, chemical and biological characteristics of the substance in question.

The preclinical, and especially, the toxicological tests are backed by stability tests on both the active ingredient and the formulations. When planning these stability tests (type I), equal attention is paid to all factors known to affect

Tab. 1 Stability Testing of New Drug Products (SOP Hoechst AG)

	Preclinical study		Clinical trial	Approval/Launch	
	Type I Preformulation	Type II Preliminary study	Type III Official study	Type IV On-going study	Type V Follow-up study
Testing material	Laboratory batches	Laboratory/Pilot batches	Pilot/Production batches	Production batches	Production batches
Objective	• Investigation of stability-influencing factors e.g. – temperature – humidity – light – oxygen – pH on the drug **substance** • Compatibility with excipients • Development of stability-specific control tests for the drug **substance** • Identification of degradation products	• Investigation of stability-influencing factors e. g. – temperature – humidity – light – manufacture on the drug **product** • Interactions with packaging materials • Identification of degradation products • Establishment of stability-relevant control tests for the drug **product** • Estimation of provisional expiration date for clinical trial samples	• Assessment of stability of the **final formulation** in the **proposed market container** for **all climate zones** • In use-tests • Confirmation of analytical methods • Establishment of – overages – specifications – expiration date – storage directions – in use-periods	• Confirmation of stability profile on the **first production batches** • Assessment of stability profile after change in – formulation – container – manufacture – synthesis route	• Monitoring of stability and specifications of retention samples
Performance	Accelerated study	Accelerated study	Long-term study Accelerated study	Comparative study Long-term study	Long-term study
Application	Basic Documentation	Basic Documentation	Marketing authorization	Amendments	Reregistration GMP Inspection

Tab. 2 Stability Testing of New Drug Products (SOP Hoechst AG)

	Type I Preformulation	Type II Preliminary study	Type III Official study	Type IV On-going study	Type V Follow-up study
Testing material	Laboratory batches	Laboratory/Pilot batches	Pilot/Production batches	Production batches	Production batches
Objective	• Investigation of stability-influencing factors e.g. – temperature – humidity – light – oxygen – pH • Development of stability specific control tests • Identification of degradation products	• Assessment of stability along with optimization of synthesis route • Interactions with packaging materials • Identification of degradation products • Establishment of stability-relevant control tests • Estimation of provisional expiration date	• Assessment of stability of the drug substance produced according to the **final synthesis route** in the **packaging similar to the packaging system proposed for storage and supply** • Confirmation of analytical methods • Establishment of – specifications – expiration date – storage directions	• Confirmation of stability profile on the **first production batches** • Assessment of stability profile after change in synthesis route	• Monitoring of stability and specifications of retention samples
Performance	Accelerated study	Accelerated study	Long-term study Accelerated study	Comparative study Long-term study	Long-term study
Application	Basic Documentation	Basic Documentation	Marketing authorization	Amendments	Reregistration GMP Inspection

Table 3: Physical, Chemical and Biological Characteristics of Active Ingredient

Physical Characteristics
• Crystallinity • Polymorphism • Particle shape and size • Specific surface area • Adsorption (i. e. water vapour) • Rate of dissolution • Melting characteristics
Chemical Characteristics
• Identity (Evidence of structure, stereochemical characteristics) • Purity (Qualitative and quantitative determination of synthesis by-products and degradation products) • Content
Biological Characteristics
• Biological activity (potency)

stability. Knowledge of the stability-influencing factors is not only of essential importance for the storage of the active ingredient but also for the galenic development of the drug product. Type I stability tests on the active ingredient are planned very carefully so as to ensure that the following objectives are attained:

- Investigation of stability-influencing factors such as temperature, humidity, light, oxygen and pH.
- The setting of preliminary processing periods for the recommended storage condition.
- Identification of degradation products.
- Preparation of documentation in order to obtain permission to perform clinical studies.

When carried out according to the plan given in Table 4, type I stability tests (active ingredient) provide information on factors such as temperature, humidity and light. If the drug substance under investigation is extremely sensitive to moisture, it is possible to include further conditions, e. g. 45%, 60%, 75% RH at a constant temperature (e. g. +25 °C). For testing the influence on light, samples are taken at the prescribed test intervals (hours or days) so that the formation of degradation products can be followed analytically.

The influence of the pH is assessed by establishing a pH stability profile.

Type I stability tests on the drug product are carried out on mixtures of the active ingredient and excipients or even on provisional preparations according to the plan given in Table 4.

Table 4: Test Plan for Stability Tests, Type I (Framework)

Storage Conditions	Test Interval (Months)							Primary Packaging
	1	3	6	12	18	24	36	
− 15°C	x	x	x	x	(x)	(x)	(x)	A
+ 5°C	(x)	x	x	x	(x)	(x)	(x)	A
+ 25°C	(x)	x	x	x	(x)	(x)	(x)	A
+ 40°C	x	x	x	(x)				A
+ 60°C	x	(x)						A
+ 40°C/75% RH	x	x	x					B
Artifical sunlight	1,	2,	3,	14 days				C

A: Inert, closed container
B: Inert, open container
C: Petri dish
(): Test carried out when necessary

It is essential to manufacture reference standards of synthesis by-products and degradation products for use in chemical and physical characterisation tests or toxicological studies.

3. Stability Tests, Type II

Type II stability tests (Tables 1 and 2) aim to:

- assess stability after optimisation of the method of manfacture (active ingredient and drug product),
- identify potential degradation products (active ingredient and drug product),
- assist in the selection of the optimal composition (drug product),
- assist in the selection of the most suitable primary packaging (active ingredient and drug product),
- determine the factors affecting stability, e.g. temperature, humidity, light and oxygen (as a complement to type I stability test),
- optimise analytical methods used to assess the quality of the product
- assist in the setting of a preliminary processing period (active ingredient)/ shelf-life (drug product) for clinical testing,
- provide data for basic documentation to be submitted in order permission to perform clinical trials.

The tests are basically carried out according to the plan given in Table 4. The length of the test depends on the shelf-life of the drug product and the length of the clinical trial.

4. Stability Tests, Type III

Whereas the samples used in type I and II stability tests are normally produced on a laboratory scale, type III stability tests are carried out using batches produced on a pilot plant or production scale. Type III stability tests (active ingredient and drug product) are performed for registration purposes in accordance with international requirements. Type III tests are carried out once the following have been established (Tables 1 and 2):

- the manufacturing procedure (active ingredient and drug product),
- the composition (drug product),
- the primary packaging material (active ingredient and drug product),
- the test criteria and control tests (active ingredient and drug product).

The test design of type III stability tests is elaborated on the basis of the results obtained with previous tests.

The alignment of international quality standards requires the pharmaceutical manufacturer to set expiry dating periods for the drug product and processing periods for the active ingredient as well as storage conditions for both. This information, which is required for all drug products intended for international markets, is based on the results obtained in the stability tests. The tests themselves are obliged to conform to international requirements, the most important of which are the guidelines issued by the EC, the USA and Japan.

Although the above guidelines are all basically very similar, there are still a few differences with regard to the prescribed storage conditions (temperature and humidity), test intervals, number of parallel tests and the evaluation and documentation of findings.

A concept for harmonising storage conditions worldwide is presented in an EC stability guideline (Table 5). The concept states that long-term studies under normal conditions should take place at +25 °C/60% RH if the phar-

Table 5: International Climate Zones and Climate Conditions Stability Guidelines, Europe (CPMP)

Climate Condition	Zone I Temperature	Zone II Mediterranean (sub-tropical)	Zone III Hot/dry or hot/moderate RH	Zone IV Very hot/humid
Mean Annual	< 20.5 °C	20.5–24 °C	> 24 °C	> 24 °C
Kinetic Mean Temperature (Virtual Temperature)	21 °C	26 °C	31 °C	31 °C
Mean Annual Relative Humidity	45%	60%	40%	70%

Table 6: Storage Conditions for Long-term Stability Testing

Climate Zone	Storage Conditions Temperature	Humidity	Testing intervals (Months) 1	3	6	9	12	18	24	30	36	48	60
(I)	+ 21 °C	45% RH	(x)	x	x	(x)	x	x	x	(x)	x	(x)	(x)
II	+ 25 °C	60% RH	(x)	x	x	(x)	x	x	x	(x)	x	(x)	(x)
(III)	+ 30 °C	40% RH	(x)	x	x	(x)	x	x	x	(x)	x	(x)	(x)
IV	+ 30 °C	70% RH	(x)	x	x	(x)	x	x	x	(x)	x	(x)	(x)
	+ 5 °C*		(x)	x	x	(x)	x	x	x	(x)	x	(x)	(x)
	+ 15 °C*		(x)	x	x	(x)	x	x	x	(x)	x	(x)	(x)

* For heat-sensitive products with corresponding storage directions.

maceutical product is intended to be stored and marketed in zones I and II; however, should the product be intended for zones III and IV (Tables 5 and 6), tests should be carried out at + 30 °C/70% RH. Long-term studies conducted at + 25 °C/60% RH are to be carried out primarily for products intended for EC countries, the USA and Japan. In contrast, long-term studies at + 30 °C/70% RH are to be carried out in order to obtain regulatory approval in other countries.

Other storage conditions, e.g., + 5 °C and + 15 °C, may be included in type III stability tests if the drug product is very sensitive to heat (Table 6).

In view of the regulations for accelerated studies in EC countries, the USA and Japan, the storage conditions for type III tests, e.g. + 40 °C (elevated temperature only) and ± 40 °C/75% RH (exposure to elevated temperature and humidity) can be regarded as internationally accepted (Table 7). In order to

Table 7: Storage Conditions for Accelerated Stability Testing

	Requirements of Stability Guidelines: Japan (MHW) 1980	USA (FDA 1984)	Europe (CPMP 1988)	Proposal for International Studies	
Humidity	+ 40 °C/ 75% RH (Acceleration test standard)	+ 40 °C/ 75% RH	+ 40 °C/ 75% RH	+ 40 °C/ 75% RH	3, 6, 12 months
Temperature	Not specified (Severe test)	Not specified	Not specified	+ 40 °C	3, 6, 12 months
Light	Not specified (Severe test)	Not specified	Not specified	Artificial sunlight	1, 2, 3 days
				Indoor light	1, 3, 6 months

Table 8: Test Items for Stability Testing (Drug Products)

Characteristics Physical	Chemical	Microbiological
• Organoleptic properties	• Content of active ingredient	• Sterility
• Properties specific to the dosage form – Hardness – In vitrodissolution – Particle size – Polymorphism – Viscosity – Moisture content, etc.	• Decomposition products • Content of other components – Antimicrobial preservatives – Antioxidants	• Bacterial endotoxins • Microbial contamination • Preservative efficacy

Table 9: Documentation

- Title (product name, dosage)
- Objective of stability study
- Summary of results with recommendations for storage conditions and shelf-lives before and after opening
- Information pertaining to batches tested
  - Number of batches
  - Composition
  - Date of manufacture
  - Batch size
- Description of packaging material
  - Containers
  - Closures for containers
- Stress tests
  - Storage conditions
  - Test intervals
- Control tests, validation, reference substances
- Tables of the test results
- Stability tests after opening/reconstitution
- Evaluation of results
  - Comments on organoleptic, physicochemical, chemical and microbiological characteristics of product
  - Explanation of shelf-life specification
  - Reasons for overage of active ingredient
  - Assessment of interactions with packaging material
  - Explanation of specifications before and after operning and of specifications for storage conditions

assess the effect of light, it is advisable to carry out tests using both artificial sunlight and indoor light.

Sampling is carried out at 3-month intervals during the first year, at 6-month intervals during the second year and once a year thereafter. Samples are taken more frequently from drug products which are expected to degrade rapidly.

If additional stability data pertaining to accelerated conditions are required, e.g. exposure to light, oxygen, temperatures of $<25\,°C$ and $>40\,°C$, relative humidity of $<60\%$ and $>70\%$, reference is made to earlier stability studies (Type I and Type II respectively).

The test items required for type III stability tests vary according to the type of drug product; an example is given in Table 8.

The results of type III stability tests are documented as shown in Table 9 as they can then be used to obtain regulatory approval for the product worldwide.

5. Stability Tests, Type IV

Type IV stability tests (Tables 1 and 2) aim to

- confirm that the stability characteristics of the prototype (registration documentation) are concordant with those of production-scale batches (active ingredient and drug product).
- Determine stability when modifications have to be made to, for example,
  - the manufacturing procedure (active ingredient or drug product),
  - the composition (drug product),
  - the primary packaging material (active ingredient or drug product).

The design of type IV stability tests on production-scale batches is based on long-term type III stability tests. Studies such as these, also known as on-going studies, have been performed according to GMP requirements for a long time.

If minor differences exist between the prototype and the finished drug product, comparative, long-term and accelerated stability tests (Tables 6 and 7) are recommended in order to determine, i.e. rule out, any possible adverse effects under elevated temperatures and extreme humidity caused by the modifications.

The documented results of comparative stability tests are usually presented to the registration authority when announcing a change to the product.

6. Stability Tests, Type V

Type V stability studies (follow-up studies) are used (see Tables 1 and 2)

- in the regular assessment of the stability of the drug substances and drug products throughout their entire commercial availability,

- to ensure ("monitor") compliance with the specifications throughout the entire shelf-life.

The samples to be examined are kept under storage conditions identical to those under which the product is intended to be marketed. These data are generated by the pharmaceutical manufacturer for quality assurance and registration purposes.

The helpful contributions of Dr. Inge Kern are gratefully acknowledged.

12 The tripartite guideline for the stability testing of new drug substances and products (draft number 3)

T. Beaumont, GB-Dartford

Introduction

Preamble

The following guideline sets out the stability testing requirement for a Registration Application within the three areas of the EC, Japan and the USA. It does not seek necessarily to cover the testing that may be required for registration in or export to other areas of the world.

The guideline seeks to exemplify the core stability data package required for new drug substances and products. It is not always necessary to follow this when there are scientifically, justifiable reasons for using alternative approaches.

The guideline provides a general indication on the requirements for stability testing, but leaves sufficient flexibility to encompass the variety of different practical situations demanded with regards to specific situations and characteristics of the materials being evaluated.

The principle that information on stability generated in any one of the three areas of the EC, Japan and the USA would be mutually acceptable in both of the other two areas has been established.

Further details of the specific requirements for sampling, statistical analysis, test requirements for particular dosage forms/packaging etc., may be found in the relevant national/regional guidelines, but the core stability package is that found in the following guideline.

Objective

The purpose of stability studies is to provide evidence how the quality of a drug substance or drug product varies with time under the influence of a variety of environmental factors such as temperature, humidity and light, and enables recommended storage conditions, re-test periods and shelf lives to be established.

Scope

The guideline primarily addresses the information required in Registration Applications for stable new drug substances and products. In case of less stable materials appropriate modifications could be necessary to cover the specific characteristics encountered. The guideline does not currently seek to cover the information required for abridged applications, variations, biotechnology/ biological products, clinical trials applications etc.

The basis for the scientific principles covering derivation of the standardised temperature and humidity conditions recommended is given in Annex 1. The choice of the test conditions defined in this guideline is made from those stated in Annex 1 for the three areas of the EC, Japan and USA.

Glossary and Information

The following terms have been in general use and the following definitions are provided to facilitate interpretation of the guideline.

Accelerated Testing

Studies designed to increase the rate of chemical degradation or physical change of a drug substance or drug product by using exaggerated storage conditions as part of the formal, definitive, storage programme. The test conditions are also used to provide comparative evidence in the short term of the equivalence of materials from various sources such as made by different manufacturers or processes or procedures or in different packagings, or where volumes or strengths of drug products are involved.

These data may also be used to assess longer term chemical effects at non-accelerated conditions and to evaluate the impact of short term excursions outside the label storage conditions such as might occur during shipping. Results from accelerated testing studies are not always predictive of physical changes.

Studies may be undertaken to identify the most sensitive dosage strength, container/closure system, and/or container size among various forms of the same drug product. For Japan the Long Term Testing could then be performed with only the selected, container/closure system, and/or container size. In this case other forms should be more stable and need therefore not be tested for the longterm testing, thus offering an alternative to Matrixing and Bracketing to reduce the number of samples tested.

Active Substance; Active Principle; Drug Substance; Medicinal Substance

The unformulated drug substance which may be subsequently formulated with excipients to produce the drug product.

Bracketing

The design of a stability schedule so that at any time point only the samples on the extremes, for example of container size and/or dosage strengths, are tested. The design assumes that the stability of the intermediate condition samples are represented by those at the extremes.

Climatic Zones

The concept of dividing the world into four zones based on defining the prevalent annual climatic conditions (see Annex 1).

Drug Product; Finished Product; Dosage Form

A pharmaceutical product type, for example tablet, capsule, solution, cream etc. that contains a drug ingredient generally, but not necessarily, in association with excipients.

Excipient

Anything other than the drug substance in the dosage form.

Expiry/Expiration Date

The date placed on the immediate container label of a drug product designating the time during which a batch of the product is expected to remain within specification if stored under defined conditions, and after which it must not be used.

Formal (Systematic) Studies

Formal systematic studies are those undertaken to a stability protocol which embraces the principles of these guidelines.

Long Term (Real Time) Testing

Stability evaluation of the physical, chemical, biological and microbiological characteristics of a drug product and a drug substance, covering the expected duration of the shelf life and retest period, which are claimed in the submission and will appear on the labelling.

Mass Balance; Material Balance

The process of adding together the assay value and levels of degradation products to see how closely these add up to 100 per cent of the initial value, with due consideration of the margin of analytical precision.

This concept is a useful scientific guide for evaluating data but it is only occasionally achievable in most circumstances. The focus may instead be on assuring the specificity of the assay, the completeness of the investigation of routes of degradation, and the use, if necessary, of identified degradants as indicators of the extent of degradation via particular mechanisms.

Matrixing

The statistical design of a stability schedule so that only a fraction of the total number of samples are tested at any specified sampling point. At a subsequent sampling point, different sets of samples of the total number would be tested. The design assumes that the stability of the samples tested represents the stability of all samples. The differences in the samples for the same drug product should be identified as, for example, covering different container/ closure systems, different batches, different strengths, different sizes etc.

Matrixing can cover reduced testing when more than one variable is being evaluated. Thus the design of the matrix will be dictated by the factors needing to be covered and evaluated. This potential complexity inclusion of specific details and examples, and it may be desirable to discuss design in advance where there is a possibility for such a discussion. In every case it is essential that all batches are tested initially and at the end of the long term testing.

Mean Kinetic Temperature

When establishing the mean value of the temperature the formula of J.D. Haynes (J. Pharm. Sci. **60**, 927–929, 1971) can be used to calculate the mean kinetic temperature. It is higher than the arithmetic mean temperature and takes into account the Arrhenius equation from which Haynes derived his formula.

Pilot Plant Scale

The manufacture of either drug substance or drug product by a procedure fully representative of and simulating that to be applied on a full manufacturing scale.

For oral solid dosage forms this is generally taken to be at a minimum scale of one tenth that of full production or 100000 dosing units for the drug product whichever is the larger. The alternative use of smaller scale batches in providing stability information needs to be justified.

Primary Stability Data

Data on the drug substance stored in the proposed packaging under storage conditions that support the proposed re-test date.

Data on the drug product stored in the proposed container-closure for marketing under storage conditions that support the proposed shelf life.

Re-Test Date

The date when samples of the drug substance should be re-examined to ensure that material is still suitable for use.

Re-Test Period

The period of time during which the drug substance can be considered to remain within the specification and therefore acceptable for use in the manufacture of a given drug product, provided that it has been stored under the defined conditions; after this period, the batch should be retested before use.

Shelf-life

The time interval that a drug product is expected to remain within the approved specification provided that it is stored under the conditions defined on the label in the proposed containers and closure. This may be regarded as equivalent to the Expiration Dating period.

Specification – Release

The combination of physical, chemical, biological and microbiological test requirements that determine a drug product is suitable for release at the time of its manafacture.

Specification – Check/Shelf-life

The combination of physical, chemical, biological and microbiological test requirements that a drug substance must meet up to its re-test date or a drug product must meet throughout its shelf life.

Storage Conditions Tolerances

The acceptable variation in temperature and relative humidity of storage facilities.

The mean value for all measured data within three months should be within specifications
Temperature: ± 2 °C
Refrigerator: 5 °C ± 3 °C
Relative humidity: ± 5%

Stress Testing

The studies undertaken, primarily on the drug substance but also on the drug product, to elucidate intrinsic stability characteristics. Such testing is part of the development strategy and is normally carried out under more severe conditions than those used for accelerated tests.

Stress testing is conducted to provide data on forced decomposition products and decomposition mechanisms for the drug substance. The severe conditions that may be encountered during distribution can be covered by stress testing of definitive batches of drug substance and drug product.

These studies should establish the inherent stability characteristics of the molecule, such as the degradation pathways, and lead to identification of degradation products and hence support the suitability of the proposed analytical procedures. The detailed nature of the studies will depend on the individual drug substance and type of drug product.

This testing is likely to be carried out on a single batch of material and to include the effect of elevated temperatures, susceptibility to hydrolyse across a wide range of pH values in solution or suspension, oxidation and photolysis on the drug substance.

Results from these studies will form an integral part of the information provided to regulatory authorities.

Light testing should be an integral part of stress testing. [The standard conditions for light testing are still under discussion and will be distributed in a further ICH document].

It is recognised that some degradation pathways can be complex and that under forcing conditions decomposition products may be observed which are unlikely to be formed under accelerated or long term testing. This information may be useful in developing and validating suitable analytical methods, but it may not always be necessary to examine specifically for all degradation products, if it has been demonstrated that in practice these are not formed.

Supporting Stability Data

Data other than primary stability data, such as stability data on early synthetic route batches of drug substance, small scale batches of materials, investigational formulations not proposed for marketing, related formulations, product presented in containers other than those proposed for marketing, information regarding test results on containers, and other scientific rationale that support the analytical procedures, the proposed re-test period or shelf life and storage conditions.

Drug Substance

General

Information on the stability of the drug substance is an integral part of the systematic approach to stability evaluation. This can involve stressing the drug substance under a variety of conditions of temperature, humidity, exposure to light, together with testing in solution or suspension at a range of high and low pH values and evaluating the influence of oxygen. This stress testing helps to determine the intrinsic stability of the molecule by establishing degradation pathways in order to identify the likely degradation products and to validate the stability indicating power of the analytical methods used.

Formal systematic studies will show that the drug substance will remain within specification during the re-test period if stored under recommended storage conditions.

Selection of Batches

Stability information from accelerated and long term testing is to be provided on at least three batches. The long term testing should cover 12 months duration on at least two of the batches at the time of submission.

The batches manufactured to a minimum of pilot plant scale should be by a synthetic route, or other appropriate method of manufacture, and procedure that simulate the final process to be used on a manufacturing scale. The overall quality of the batches of drug substance should be representative of the quality of material made on a manufacturing scale.

Data on laboratory scale batches may be provided to justify absence of 12 months definitive data on three pilot plant or larger scale batches. If not covered by the information provided in the Registration Application confirmatory stability information should be provided at a later stage.

Test Methods and Test Criteria

The testing should cover those features susceptible to change during storage and likely to influence quality, safety or efficacy. Validated stability-indicating testing methods must be applied. The need for the extent of replication will depend on the results of validation studies.

Specification

Limits of acceptability should be derived from all the available stability information. It may need to include specific limits for individual and total degradation products, the justification for which could be influenced by the level observed in material used in pre-clinical studies.

Storage Test Conditions

The length of the studies and the storage test conditions should be sufficient to cover storage, shipment and subsequent use. Application of the same storage conditions as applied to the drug product will facilitate comparative review and assessment. Other storage conditions are allowable if justified. In particular, temperature sensitive drug substances should be stored under an alternative, lower temperature condition which will then become the designated long term testing storage temperature. The six months accelerated testing should then be carried out at a temperature at least 15 °C above this designated long term storage temperature (together with the appropriate relative humidity conditions for that temperature). The designated long term testing conditions will be reflected in the labelling and re-test date.

Conditions		**Minimum Time Period at Submission**
Long term testing	25 °C ± 2 °C/NLT 60% RH ± 5%	12 Months
Accelerated Testing	40 °C ± 2 °C/75% RH ± 5%	6 Months

NLT = Not less than.

Should any significant change occur under the conditions used for accelerated testing it is strongly recommended that additional testing be carried out over at least a six months period at an intermediate condition (such as 30 °C ±

2 °C/NLT 60 per cent RH ±5 per cent) providing a more severe thermal challenge than the designated long term testing condition, and that this information should be included in the Registration Application.

The long term testing will be continued for a sufficient period of time beyond 12 months to cover all appropriate re-test periods, and the further accumulated data can be submitted to the Authorities during the assessment period of the Registration Application.

Testing Frequency

Frequency of testing should be sufficient to establish the stability characteristics of the drug substance. Testing under the defined long term conditions will normally be every three months, over the first year, every six months over the second year and than annually. Reduced testing can be acceptable if justified, especially for stable substances, shown to be stable by stress testing.

Packaging/Containers

The containers to be used in the long term, real time stability evaluation should be the same as or simulate the actual packaging used for storage and distribution.

Evaluation

A systematic approach should be adopted in the presentation and evaluation of the stability information, which should cover as necessary physical, chemical, biological and microbiological test characteristics. Where applicable and meaningful, for example where there has been a significant change, formal statistical evaluation such as regression analysis may be applied. Any such evaluation should consider not only the assay but also the levels of degradation products and other appropriate attributes.

The drug substance must meet the requirements of the specification throughout its re-test period. Extrapolation may be acceptable where supported by accelerated and real time data. Accelerated data may be particularly useful in comparing different manufacturers, scales of manufacture, variables of manufacturing process and different batches to establish batch to batch variability before real time data are available.

Statements

A re-test period should be derived from the stability information.

Labelling

Based on the stability evaluation the drug substance may be assigned where suitable and appropriate to one of the following categories and labelled accordingly.

1. No specific storage precautions conditions necessary.
2. Store at up to 30 °C.
3. Store at up to 25 °C.
4. Store at 2–8 °C – refrigerate; do not freeze.
5. Store below 8 °C.
6. Store below – 5 °C – freeze.
7. Store below – 18 °C – deep freeze.

A storage temperature range may be used, and should specifically be stated where required in particular for drug substances that cannot tolerate freezing.

Reference to storage under ambient conditions or room temperature is unacceptable. Alternative presentations such as "Store below 25 °C" may be acceptable. General precautionary statements of "Protect from Light" and "Store in a Dry Place" may be included.

Drug product

General

The design of the stability programme for the finished drug product should be based on the knowledge of the behaviour and properties of the drug substance and the experience gained from the stability studies on the drug substance. The likely changes on storage and the rationale for the selection of product variables to include in the testing programme should be stated.

Selection of Batches

Stability information from accelerated and long term testing is to be provided on at least three batches of the same formulation proposed for marketing. The long term testing should cover at least 12 months duration on at least two of the batches at the time of submission. The manufacturing process to be used should meaningfully simulate that which would be applied to large scale batches for marketing. The process should provide product of the same quality intended for marketing, and meeting the same analytical specification as to be applied for release of material. Where possible, batches of the finished product should be manufactured using identifiably different batches of drug substance.

Data on laboratory scale batches may be provided to justify absence of 12 months definitive data on three pilot plant or larger scale batches. If not covered by the information provided in the Registration Application confirmatory stability information should be provided at a later stage on three early production batches manufactured for marketing, normally the first three consecutive production batches.

When justified, data on associated formulations or packaging may be cross-referenced as supportive information if long term definitive data are not yet available.

Test Methods and Test Criteria

The testing should cover those features susceptible to change during storage and likely to influence quality, safety or efficacy. Analytical test methods should be fully validated and the assays should be stability indicating. The need for the extent of replication will depend on the results of validation studies.

The range of testing should cover not only chemical stability but also effectiveness of additives such as preservative, physical properties and characteristics and organoleptic properties, where feasible.

Specifications

Limits of acceptance should relate to the release limits (where applicable), and be derived from consideration of all of the available stability information. The shelf life specification could allow acceptable and justified deviations from the release specification based on the stability evaluation and the changes observed on storage. It may need to include specific limits for degradation products, the justification for which could be influenced by the levels observed in material used in pre-clinical studies, where applicable. The justification for the limits proposed for certain other tests such as particle size or dissolution rate may require reference to the results observed for batch(es) used in biovailability and/or clinical studies.

Storage Test Conditions

For drug products involving a new drug substance a minimum of 12 months stability information under the intended, long term, storage test condition should be provided, together with an assurance that testing will continue to cover the expected shelf life. Where available supporting stability data should be supplied.

Other storage conditions are allowable. In particular temperature sensitive drug products should be stored under an alternative lower temperature condition, which will then become the designated long term testing storage temperature. The six months accelerated testing should then be carried out at a temperature at least 15 °C above the designated long term storage temperature (together with the appropriate relative humidity conditions for that temperature). The designated long term testing conditions will be reflected in the labelling and expiration date.

Storage under conditions of high relative humidities applies particularly to solid dosage forms in semi-permeable packaging. For products such as solutions, suspensions etc., contained in packs designed to provide a permanent barrier to water loss, specific storage under conditions of high relative humidity is not necessary but the same range of temperatures should be applied. Low relative humidity can adversely affect products packed in semi-permeable containers (e.g. solutions in plastic bags, nose drops in small plastic container etc.) and consideration should be given to appropriate testing.

Conditions		**Minimum Time Period at Submission**
Long term testing	25 °C ± 2 °C/NLT 60% RH ± 5%	12 Months
Accelerated Testing	40 °C ± 2 °C/75% RH ± 5%	6 Months

NLT = Not less than.

For products which are intended to soften or melt at or below 37 °C (such as suppositories and ophthalmic ointments), or where significant changes occur under the conditions used for accelerated testing, it is strongly recommended that additional testing be carried out over at least a six months period at an intermediate condition (such as 30 °C ± 2 °C/NLT 60 per cent RH ± 5 per cent), thus providing a more severe thermal challenge than the designated long term testing condition. This information should be included in the Registration Application.

The long term testing will be continued for a sufficient period of time beyond 12 months to cover the proposed shelf life, and the further accumulated data can be submitted to the Authorities during the assessment period of the Registration Application.

The long term testing should cover the full expiration dating period and the labelling will reflect the results of the test under the conditions chosen for the long term testing.

Testing Frequency

Frequency of testing should be sufficient to establish the stability characteristics of the drug product. Testing will normally be every three months over the first year, every six months over the second year and then annually. Reduced testing can be acceptable if justified, especially for stable products.

The use of matrixing or bracketing can be applied, if justified.

Packaging Materials

The testing should be carried out in the final packaging proposed for marketing or, when justified, in packaging that simulates the final one. Additional testing of unprotected drug product can form a useful part of the stress testing and pack evaluation, as can studies carried out in other related packaging materials in supporting the definitive pack(s).

Evaluation

A systematic approach should be adopted in the presentation and evaluation of the stability information, which should cover as necessary physical, chemical, biological, microbiological test characteristics, including particular properties of the dosage form (for example dissolution rate for oral dosage forms). Where applicable and meaningful, for example where there has been a significant change, formal statistical evaluation such as regression analysis may be applied. Any such evaluation should consider not only the assay but also the levels of degradation products and other appropriate attributes. Where appropriate attention may be paid to reviewing the adequacy of the mass balance obtained.

Extrapolation of the shelf-life may be acceptable, where supported by accelerated and real time data. Accelerated data may be particularly useful in comparing different manufacturers, scales of manufacture, variables of manufacturing process and different batches to establish batch to batch variability before real time data are available.

An explanation should be provided if individual batches have a significantly different stability and degradation performance.

The stability of the drug product after reconstituting, further dilution either in specifically proposed mixtures or where a specific claim is made, should be addressed to provide appropriate information of storage and handling.

Statements/Labelling

Based on the stability evaluation the drug product may be assigned where suitable and appropriate to one of the following categories and labelled accordingly.

1. No specific storage conditions necessary*.
2. Store at up to 30 °C.
3. Store at up to 25 °C.
4. Store at 2–8 °C – refrigerate; do not freeze.
5. Store below 8 °C.
6. Store below − 5 °C – freeze.
7. Storc below − 18 °C – deep freeze.

A storage temperature range may be used and should specifically be stated where required, in particular for drug products that cannot tolerate freezing.

Reference to storage under ambient conditions or room temperature is unacceptable. Alternative presentations such as "Store below 25 °C" may be acceptable. General precautionary statements of "Protect from Light" and "Store in a Dry Place" may be included, but should not be used to cover products with stability problems. Instead a packaging material should be used which offers the product suitable protection.

There should be direct linkage between the label statement and the demonstrated stability characteristics of the drug product.

* Up to now the USA requires statements on storage conditions for all products.

Annex 1

Climatic Zones

For world-wide stability testing, the world is divided into four climatic zones (Futscher N., Schumacher P, Phar. Ind., **34**, 479–483, 1972) for which the following storage conditions have been established on the basis of the values measured for a number of cities in the relevant climatic zone. Further information on the climatic data measured in the open and in storage rooms, and the derivation of the final combination of temperature and relative humidity covering the four climatic zones may be found in the following publications (Grimm W., Drugs made in Germany, **28**, 196–202, 1985, and **29**, 39–47, 1986). Climatic Zone II is established as being prevalent in covering EC, Japan and the USA.

Climatic Zone	Measured Data in the Open			Measured Data in Storage Room		
	[°C]	[% R.H]	[mbar]	[°C]	[% R.H]	[mbar]
I temperate	10,9	75	9,9	18,7	45	9,8
II Mediterranean Sub Tropical	17,0	70	13,5	21,1	52	13,1
III hot dry	24,2	39	11,9	26,0	54	18,0
IV hot humid	26,5	77	26,6	28,4	70	27,0

Climatic Zone	Calculated Data <19 °C = 19 °C*				Derived Data		
	[°C]	[°C MKT]**	[%]	[mbar]	[°C]	[%]	[mbar]
I temperate	20,0	20,0	42	9,9	21	45	11,2
II Mediterranean Sub Tropical	21,6	22,0	52	13,5	25	60	19,0
III hot dry	26,4	27,9	35	11,9	30	35	15,0
IV hot humid	26,7	27,4	76	26,6	30	70	30,0

* Drug products are stored in rooms. Consequently, 19 °C is used as the temperature for all months during which the temperatures measured are below 19 °C

** Mean Kinetic Temperature

13 General concept for stability testing

W. Grimm, D-Biberach/Riß

The aim of stability testing is to ensure that the full efficacy of a drug product can be guaranteed until the end of the expiration date.

The realisation of this aim is made possible by a general concept based on the 12 principles for stability testing [1].

The 12 principles are as follows:

- □ systematic development
- □ selection of batches and samples
- □ test criteria
- □ analytical methods
- □ tolerances
- □ storage conditions
- □ testing intervals
- □ storage period
- □ number of batches
- □ packaging materials
- □ evaluation
- □ statements

The stability test based on these principles

- □ is generally applicable
- □ allows statements to be made concerning stability
  - for a country
  - for a climatic zone
  - for worldwide introduction
- □ can be applied to the problem under consideration
  - new active ingredient, new drug product
  - new dosage form
  - new dosage
  - major change.

Primarily, however, a stability programme is presented for climatic zone II which covers the most important pharmaceutical markets EC, Japan and the USA.

Stability information is the product of thorough investigations and is derived from a large quantity of generated data. The basis for this is systematic development in a logically structured and optimally coordinated system. By this means the state of knowledge can be continuously extended and deepened.

Stability testing for a new drug product can be divided into six steps [2]:

- ☐ tests with the active ingredient
- ☐ preformulation and formulation finding for the clinical samples and the final dosage form
- ☐ acceleration tests with selected formulations and packaging materials
- ☐ long-term tests
- ☐ follow-up tests
- ☐ tests in the event of changes during continuous production.

This division into six steps makes it possible

- ☐ to define the problem under consideration,
- ☐ to establish the required level of analytical effort,
- ☐ to estimate the capacity,
- ☐ to use the results for subsequent development.

These six steps assume that a new drug product is to be developed for a new active ingredient.

If a new dosage form or a different dosage subsequently has to be developed, these activities can be based on all the results already available.

The relevant stage of development is then entered. This allows an extremely rational procedure to be adopted. It is therefore important to

- ☐ define the objective precisely,
- ☐ prepare a schedule of tests

for each individual stage.

The individual stages of systematic development and the stability tests deriving from them will now be described.

1. Tests with the Active Ingredient

The tests are based on development chemistry, which demonstrates the following:

- ☐ proof of structure,
- ☐ profile for possible impurities,
- ☐ chemical and physical properties of the active ingredient.

The tests with the active ingredient are of a general nature and are not specific to any particular dosage form. Consequently the results are generally applicable.

1.1 Accelerated tests with the active ingredient

1.1.1 Objective

Clarification of the stability behaviour of the active ingredient with reference to physical and chemical properties.

A requirement profile is prepared for this purpose. This means:

Listing the factors that could influence the chemical and physical stability of the active ingredient during handling, processing and storage, e. g.
temperature
temperature + humidity
temperature + humidity + concentration
temperature + pH
temperature + pH + buffer concentration
temperature + 0_2
light.

Information on solubility, pH, polymorphism and chirality is also necessary.

The tests to be established on the basis of the requirement profile take into account the results obtained in development chemistry. For example, this means that the possible impurities of the active ingredient are only taken into account when optimising the analytical procedure if they are relevant to the stability, i. e. if they are potential decomposition products.

1.1.2 Schedule of Tests

Selection of batches and samples:	Representative active ingredient batches, as far as possible. However, these batches must comply with the specifications of the testing specifications in all parameters. The "impurity profile" and particle size distribution are especially critical.
Test criteria:	Appearance Assay, decomposition products, physical properties.
Analytical techniques:	Specific for stability testing, validated: selectivity, linearity, limit of quantitation, precision, ruggedness. Apparatus qualified.
Tolerances:	These are not relevant here, since the changes first have to be determined.
Storage conditions:	Storage in open containers 21 °C/45%; 25 °C/60%; 25 °C/75%, 30 °C/70%; 40 °C/75% container: 40 °C, 50 °C, 60 °C
Testing intervals:	≥ 4
Storage period:	• Until equilibrium is reached • ≤ 3 months.
Number of batches:	1
Packaging materials:	Flask with ground-glass stopper, glass container with twist-off closure.

Evaluation:	Comparison of results, statistics, reaction kinetics.
Information:	Establishment of the stability profile for the active ingredient. This contains notes on handling, re-test date, storage, possible dosage forms and expiration dates. The active ingredient stability profile forms the basis for the pharmaceutical development and must therefore always be available before its inception. The decomposition products that occur in the stress tests are recorded analytically. Further follow-up, structural elucidation and toxicity testing are only performed if the decomposition products can also occur under real conditions.

1.1.3 Practical Example

The influencing factors to be investigated are now illustrated by means of a practical example.

1.1.3.1 Temperature

Test sample:	Active ingredient in 25-ml glass container with twist-off closure, milled or with particle size distribution as used.
Test conditions:	
Test criteria:	Appearance, decomposition and assay of active ingredient.
Storage temperatures:	40 °C, 50 °C, 60 °C
Testing frequency:	0, 4, 8, 12 weeks
Storage period:	Up to 12 weeks

1.1.3.2 Temperature, Moisture

Test sample:	Active ingredient stored in open containers until equilibrium is reached at 21 °C/45%, 25 °C/60%, 25 °C/75%, 30 °C/70% and 40 °C/75%. The sample with the highest water absorption is investigated further.
Test conditions:	As in Section 1.1.3.1.

1.1.3.3 Temperature, Moisture, Active Ingredient Concentration

Test sample:	1% and 5% aqueous solution or aqueous suspension or slurry in 25-ml flask with ground-glass stopper.
Test conditions:	As in 1.1.3.1.

1.1.3.4 pH

Sample:	1% aqueous solution or slurry in Mcllvaine buffers (0.1 M citric acid, 0.2 M disodium phosphate), pH 3, 4, 5, 6, 7 in volumetric flask.
Test conditions:	
Test criteria:	Appearance, decomposition and assay of active ingredient.
Storage temperature:	60 °C
Testing frequency:	0, 4, 8, 12 weeks
Storage period:	Up to 12 weeks

1.1.3.5 pH and Buffer Concentration

Test sample:	At the optimum pH obtained from test 1.1.3.4, 1% aqueous solution or slurry in Mcllvaine buffer, double buffer concentration (0.2 M citric acid, 0.4 M disodium phosphate) in small volumetric flask.
Test conditions:	As in 1.1.3.4.

1.1.3.6 Oxidation

Test sample:	1% aqueous solution or slurry in 0.3% H_2O_2 solution in volumetric flask.
Test conditions:	
Test criteria:	Appearance, decomposition and assay of active ingredient.
Storage temperatures:	40 °C, 50 °C
Testing frequency:	0, 4, 8, 12 weeks
Storage period:	Up to 12 weeks.

1.1.3.7 Light (xenon lamp)

Test samples:	Active ingredient in colourless glass, 1% aqueous solution (or organic solvent) in colourless glass.

Test conditions:	
Test criteria:	Appearance, decomposition and assay of active ingredient
Storage temperatures:	About 10 °C above room temperature due to the Xenon lamp
Testing frequency:	8, 16, 24 hours
Storage period:	Up to 24 hours

Information on polymorphism is contained in the development chemistry report. If polymorphous forms are present, all non-dissolved samples are subjected to a DTA analysis at the end of storage.

If an enantiomer is present, testing for racemisation is performed during the course of stability studies.

The analytical methods used are validated in each case.

The results are presented as an active ingredient stability profile including the test methods used and validated.

1.2 Long-Term Tests with the Active Ingredient

These tests follow the accelerated tests and accompany the development of the formulation. The samples are stored under the conditions of the relevant climatic zone or zones.

1.2.1 Objective

- ☐ Confirmation of the results of the accelerated tests
- ☐ Clarification of the stability behaviour under storage conditions of the corresponding climatic zone or zones where the active ingredient will be stored or used.
- ☐ Determination of re-test date

1.2.2 Schedule of Tests

- ☐ Selection of batches, storage samples
  Representative active ingredient batches as far as possible. However these batches must comply with the specifications of the testing specifications in all parameters.
  At least one representative production batch should be included whenever possible.
- ☐ Test criteria
  Appearance, assay, decomposition products, physical properties
- ☐ Analytical methods
  Specific for stability testing, validated: Selectivity, linearity, limit of quanti-

tation, precision, ruggedness. Apparatus qualified. Usually the analytical methods of the accelerated tests will be applied.

- ☐ Tolerances
  Those of testing specifications
- ☐ Storage conditions
  According to the four climatic zones the following storage conditions were established for these zones:

climatic zone	storage condition
I	21 °C/45% r.H.
II	25 °C/60% r.H.
III	30 °C/35% r.H.
IV	30 °C/70% r.H.

Which and how many of these are used depends on where and how far the active ingredient will be stored, shipped and used.

- ☐ Testing intervals and storage period
  - 0, 3, 6, 9, 12, 18, 24, 36, 48, 60 months
    From the storage conditions, testing intervals and storage period, the following schemes have been derived for climatic zone II and world-wide use.

Climatic zone II

temp. [°C]	rel. hum. [%]	storage period and testing intervals [months]									
		0	3	6	9	12	18	24	36	48	60
25	60	x	x	x	x	x	x	x	x	x	x
30	70		x	x	x	x					
40	75		x	x							

Worldwide

temp. [°C]	rel. hum. [%]	storage period and testing intervals [months]									
		0	3	6	9	12	18	24	36	48	60
25	60	x	x	x	x	x	x	x	x	x	x
30	70		x	x	x	x	x	x	x	x	x
40	75		x	x							

- ☐ Number of batches
  3 batches

☐ Container
  It should be the same or simulate the actual packaging used for shipping and or storage.
  Up to now tight containers, such as glass bottles with screw cap or twist-off bottles are used.

☐ Evaluation
  The analytical results are evaluated as follows
  - Batch per storage condition:
    For each test parameter it is investigated whether a significant change has occurred compared to the initial value
  - The storage conditions of a batch
    The results obtained in the different storage conditions are compared.
  - Different batches
    The last step is the critical comparison of individual batches

☐ Stability information
  All the results and the stability information are compiled in a stability report. This contains:
  - the batch information
  - the stability testing specifications
  - the results
  - a critical assessment of the analytical technique used and of the results
  - a statement for the re-test date.

2. Preformulation and Formulation Finding for the Clinical Samples and Final Dosage Form

The tests are based on the active ingredient stability profile.

This stage covers the process of development from the first preformulation tests to the clinical samples and the potentially final formulation for the desired dosage form. Consequently, it includes the actual pharmaceutical-technological development.

Since a different dosage form is frequently used for the clinical samples than is planned for later market introduction, e.g. capsules for clinical trials, tablets for the market, a double-tracked approach is necessary in some cases. From the analytical viewpoint, this gives rise to the following objective:

2.1 Objective

☐ Dosage form related tests with the active ingredient

☐ Specification of individual excipients

☐ Investigation of external factors that have an influence on individual excipients

- □ Compatibility tests active ingredient-excipients
- □ Clarification of technological influencing factors
- □ Selection of formulation for clinical samples
- □ Selection of formulation and optimisation of the potentially final formulation

Before establishing the schedule of tests and before the practical implementation, all the results already available are evaluated. If this procedure is consistently followed, a broad spectrum of evaluable data, especially for frequently used excipients, is soon available.

The requirements placed on the different dosage forms are highly variable. Furthermore, an individual approach is adopted for each specific case which is shaped by the corporate philosophy as well as the skill and experience of the developer.

A schedule of tests can thus only be sketched in very general terms:

2.2 Schedule of Tests

Selection of batches and samples:	Active ingredient, excipients, active ingredient-excipient mixtures, dosage forms. The active ingredient and excipient batches used must comply with the specifications of the quality control testing specifications.
Test criteria:	Appearance, relevant physico-chemical parameters, active ingredient decomposition and assay.
Analytical methods:	Specific to stability testing, for comparative tests frequently semi-quantitative, e.g. HPTLC. Usually not validated. Methods for physico-chemical determinations are optimised and validated, e.g. dissolution rate.
Tolerances:	Not relevant, tests are performed to identify changes.
Storage conditions:	Standard storage conditions for stress tests and long-term tests, relevant for the dosage form. • ≥ – 10 °C, 4 °C 4–40 °C variable temperature 21 °C/45%, 25 °C/60%, 25 °C/75%, 30 °C/70%, 40 °C/75%, 40 °C, 60 °C. It is essential not to use any other storage conditions, so that these results can be built upon in the subsequent developmental stages.
Testing intervals:	Individually determined, depending on the stability behaviour.
Storage period:	Problem-orientated

- e.g. until equilibrium is reached
- not more than 3 months

Evaluation and statements:

- dosage form related active ingredient profile
- parameters and stability results of individual excipients
- selected formulations for clinical samples and potential final formulations.

2.3 Practical Examples

The tests necessary at this stage can only be carried out in close cooperation between the pharmaceutical-technological and the analytical laboratory. The tests also depend on

☐ the active ingredient, its solubility and stability,
☐ the dosage form to be developed,
☐ the experience and skill of the developer.

Only a few points may be given as examples.

2.3.1 Dosage form related active ingredient profile

This concerns all the solubility problems of an active ingredient under consideration:

☐ solubility profile in dependence on the pH, particle size . . .
☐ tests to improve the solubility or dissolution rate.

2.3.2 Stability behaviour of excipients

☐ sorption tests
☐ sensitivity to oxygen
☐ sensitivity to light
☐ sensitivity to pH

2.3.3 Selection of formulation

☐ compatibility tests active ingredient – excipients
☐ planning of tests, factorial design
☐ technological influencing factors
☐ optimisation of pH for liquid, semi-solid and, if appropriate, solid dosage forms

- ☐ oxygen and sensitivity to light-discoloration
- ☐ content uniformity
- ☐ phase separation with semi-solid forms
- ☐ solubility problems with semi-solid and liquid dosage forms
- ☐ optimised in vitro method for determination of dissolution rate

3. Stress Tests with the Selected Formulation, Selection of Packaging Materials

These tests follow directly upon step 2. This is the main area of emphasis in development analysis.

The requirement is to establish whether the formulation selected in screening in step 2 is not only stable in relative terms but also absolutely stable enough for clinical trials, for product introduction.

3.1 Objective

- ☐ optimisation and validation of analytical techniques
- ☐ identification of weaknesses of the formulation and of parameters that have a limiting effect on the shelf-life
- ☐ establishment of tolerances for tolerable changes during storage
- ☐ identification of problems that could potentially arise during storage and especially during transport
- ☐ selection of suitable packaging materials
- ☐ elucidation of mechanisms of decomposition
- ☐ establishment of stability periods for clinical samples
- ☐ establishment of provisional stability periods for CTX and IND documents
- ☐ establishment of provisional stability periods for the final dosage form.

The stress tests must be thought out and planned very carefully, since after all the aim is to make absolute statements and derive stability predictions. Particularly critical aspects are:

- ☐ storage conditions
- ☐ packaging materials.

If a stability prediction embracing all test criteria is to be prepared, a distinction must be drawn between storage for

- ☐ organoleptic and physico-chemical stability, where the laws of reaction kinetics do not apply,
- ☐ chemical (active ingredient, preservatives) stability, where the laws of reaction kinetics are applicable.

In testing for organoleptic and physico-chemical stability, therefore, the samples of solid dosage forms are stored in open containers until equilibrium is reached at the

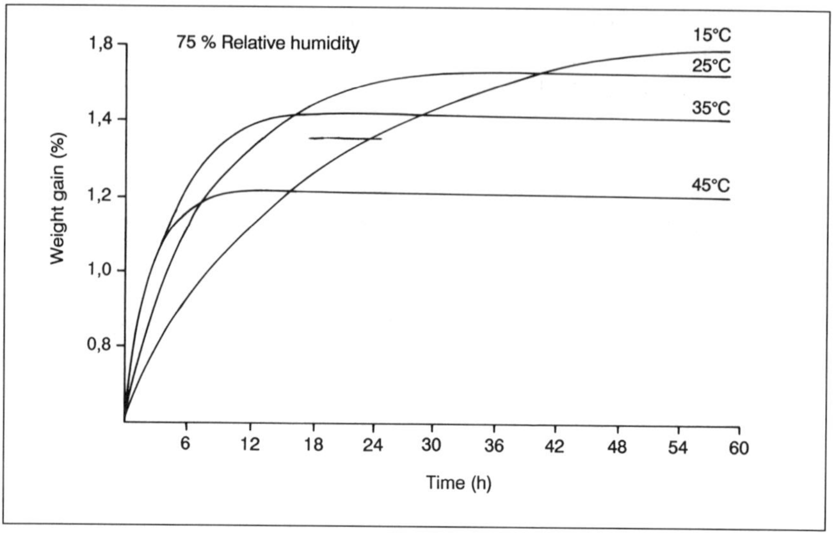

Fig. 1: Adsorption as a function of the temperature

- ☐ storage conditions of long-term testing
  21 °C/45%, 25 °C/60%, 30 °C/70%
- ☐ 25 °C/75% stress condition with high humidity
- ☐ 40 °C/75% (stress condition of EC, Japan, USA)

It is then established which test criteria have undergone changes and to what extent these changes are tolerable.

At higher temperatures, e.g. 40 °C, desorption and loss of moisture also occurs at higher relative humidities, as is shown by the following examples: figure 1, tables 1, 2 [3, 4].

These examples demonstrate the problems that arise with stress condition 40 °C/75%, especially as regards making a stability prediction. There is no logical relationship between 40 °C/75% and the stability conditions of long-

Table 1: Weight increase of tablets under various storage conditions

storage condition (°C) (% r.H.)	partial press. P_D (mbar)	water content of air (q/m³)	increase in weight (%)
16/60	10.9	8.2	2.7
21/60	14.9	11.1	2.7
26/60	20.2	14.6	2.3
31/60	26.9	19.2	2.1
27/79	25.2	20.4	3.3
31/70	31.5	22.5	2.8
41/84	65.3	45.1	2.6

Table 2: Storage condition and weight gain

storage condition	weight gain
21 °C/45%	1.8%
25 °C/60%	2.0%
25 °C/75%	3.1%
30 °C/70%	2.8%
40 °C/75%	2.6%

term testing. The examples also illustrate the problem associated with the packaging materials for stress tests. Unless packaging materials impermeable to water vapour are used for stress tests with solid dosage forms, the samples lose moisture at different rates in the temperature range 40–60 °C and the results are not suitable for a reaction kinetics calculation.

Packaging materials permeable to water vapour can however also result in a falsification of the results for semi-solid and liquid dosage forms if varying degrees of weight loss occur which lead to differences in the active ingredient concentration or ion strength.

The use of inert standard packaging materials that are impermeable to water vapour is thus an important precondition for stress tests which are to be evaluated in terms of reaction kinetics, and on the results of which stability predictions are to be based.

The schedule of tests shown therefore allows for the use of

☐ separate storage conditions
☐ standard packaging materials

3.2 Schedule of Tests

Selection of batches and samples:	• Batches if possible produced from representative batches of active ingredient and excipients. • Validated laboratory batch parallel to • Validation samples • Clinical sample • Packaged in standard packaging materials.
Test criteria:	• Active ingredient assay, content of decomposition products • Preservative assay • Organoleptic and physico-chemical, relevant to the dosage form.
Analytical methods:	Specific for stability testing, validated, apparatus qualified.
Tolerances:	Are derived from the results.

Storage conditions **Organoleptic and physico-chemical stability**	
Solid dosage forms:	Storage in open containers until equilibrium is reached (2–4 weeks) at 21 °C/45 °C r. H.; 25 °C/60% r. H.; 25 °C/75% r. H.; 30 °C/70% r. H.; 40 °C/75% r. H.
Semi-solid dosage forms:	≥ − 10 °C, temperature varying between 4 °C and 40 °C, 4 weeks, 24-hour rhythm.
Liquid dosage forms:	≥ − 10 °C, 4 weeks.
Chemical stability	40 °C, 50 °C, 60 °C, 70 °C, 80 °C number: ≥ 3 temperatures appropriate to the dosage form.
Testing intervals:	Individually determined, depending on the problem and stability behaviour. When testing for chemical and microbial stability ≥ 4 determinations including initial analysis.
Storage period:	≤ 3 months.
Number of batches:	1
Packaging materials:	Standard packaging materials impermeable to water vapour, e. g. glass containers with twist-off closure. See also 3.4, selection of packaging materials
Evaluation:	• Test for significant changes with the aid of statistics • Reaction kinetics calculation
Information:	• Establishment of test parameters for long-term testing • Fixing of tolerances, especially for changes during storage, as far as possible • Selection of suitable packaging materials • Structural elucidation of significant decomposition products. Only those decomposition products are followed up that can occur under real conditions. • Expiration dates for clinical samples • Provisional expiration dates as forecasts • Documents for CTX and IND, registration

The results represent an important aid to deciding whether the pharmaceutical-technological development can be regarded as complete and whether scaling up or long-term testing can be initiated. They also serve as supporting data for CTX and IND documents and for registration and licensing, until the expiration date has been covered by the complete storage period.

3.3 Practical Examples

The following are presented as examples:
Solid dosage forms: Tablets, Capsuls
Semi-solid dosage forms: Cream, Oinment
Liquid dosage forms: Injection solution

3.3.1 Tablets normal, Capsuls

Organoleptic and physico-chemical stability

The samples are stored in containers without closure until equilibrium is reached at: 21 °C/45%, 25 °C/60%, 25 °C/75%, 30 °C/70%, 40 °C/75%.
Test parameters:
Tablets: appearance, average weight, disintegration time, dissolution rate, hardness.
Capsuls: appearance, elasticity, average weight, average weight of filling, disintegration time, dissolution rate

Chemical stability

The untreated samples and the samples stored at 25 °C/75% until equilibrium was reached are stored at 40 °C, 50 °C, 60 °C and 70 °C for up to 12 weeks in glass containers with twist-off closure and are analysed at least four times.

Test parameters: appearance, average weight, disintegration time, active ingredient decomposition, active ingredient assay.

Photostability

The samples in open containers are irradiated with a xenon lamp for up to 24 hours.

Test parameters: appearance, active ingredient decomposition, active ingredient assay.

For the tests to be performed, a distinction is drawn between:

- ☐ clinical samples
  - a dosage or representative dosage 1
  - dosage series 2-x
- ☐ final formulation
- ☐ validation samples of the final formulation.

Schedule of tests
Organoleptic, physico-chemical stability, storage in open containers

samples	storage condition				
	21 °C/45%	25 °C/60%	25 °C/75%	30 °C/70%	40 °C/75%
clinical sample 1		x		x	
final formulation	x	x	x	x	x
validation samples		x		x	

Chemical stability
Standard packaging materials, glass containers with twist-off closure

sample	pretreatment	storage conditions			
		40 °C	50 °C	60 °C	70 °C
clinical sample 1	none	x		x	
	25 °C/60%	x		x	
clinical sample 2–x	none	x			
final formulation	none	x	x	x	x
	25 °C/75%	x	x	x	x

The validation samples are tested only for organoleptic and physico-chemical stability and are therefore not included in tests for chemical stability

3.3.2 Cream, Oinment

Organoleptic and physico-chemical stability

The samples are stored in standard tubes for up to 4 weeks at −10 °C to determine whether any changes occur which are irreversible at room temperature.

Test parameters: appearance, homogeneity, particle size (if active ingredient is in suspension), recrystallisation, consistency.

The samples in standard tubes are stored for up to 2 weeks at a temperature varying between 4 °C and 40 °C. The temperature is varied in a 24-hour rhythm.

Test parameters: appearance, homogeneity, particle size (if active ingredient in suspension), consistency.

The samples in standard tubes are stored vertically and inverted at 40 C for 12 weeks.

Test parameters: appearance, homogeneity, content uniformity.

For this purpose, the tubes are cut open and one sample from the beginning, middle and end of the tube is taken and analysed. The parameters determined are the active ingredient assay, preservative assay, active ingredient decomposition and preservative decomposition.

Chemical stability

The samples are stored at 30 °C, 40 °C, 50 °C and 60 °C for up to 12 weeks and analysed at least 4 times. Because of the phase separation that occurs at elevated temperatures, the sample quantity required for an analysis is weighed into a small volumetric flask at the beginning of each test.

Test parameters: appearance, odour, pH, active ingredient assay, preservative assay, active ingredient decomposition, preservative decomposition.

If validation samples are included in the stress tests parallel to the validated laboratory batch, they are generally tested only for organoleptic and physico-chemical stability.

3.3.3 Injection solution

Organoleptic and physico-chemical stability

The samples are stored in injection vials at − 10 °C for up to 4 weeks to determine whether any changes occur that are irreversible at room temperature.

Test parameters: appearance, clarity, pH.

Chemical stability

The samples are stored at 40 °C, 50 °C, 60 °C and 70 °C for up to 12 weeks and analysed at least 4 times. If possible in original ampoules, otherwise in flasks with ground-glass stopper.

Test parameters: appearance, clarity, pH, active ingredient decomposition, active ingredient assay.

Photostability

The samples in the original packaging are irradiated with a xenon lamp for up to 24 hours.

Test parameters: appearance, active ingredient decomposition, active ingredient assay.

Packaging material interaction

If an injection vial with a rubber stopper is intended as packaging material, the vials are stored in inverted position at 40 °C for 12 weeks.

Test parameters: appearance, clarity, pH, active ingredient decomposition, active ingredient assay, extracted substances.

In the case of an emulsion or suspension, the stress tests apply for these dosage forms.

3.4 Selection of Packaging Materials

Most of the stress tests are carried out in standard packaging materials. This makes it possible

- ☐ to make definite stability statements concerning bulkware,
- ☐ to derive standardised, comparable statements,
- ☐ to test for the influence of moisture in a controlled manner, especially for solid dosage forms.

The following standard packaging materials are used:

Solid dosage forms:	25-mal glass container with twist-off closure
Semi-solid dosage forms:	Standard tube Small volumetric flask Aluminium tube, inert internal lacquering
Liquid dosage forms:	25-ml volumetric flask

Selection of packaging materials for solid dosage forms

On the basis of the results of the stress tests for solid dosage forms the sensitivity to moisture can be determined and suitable packaging materials can be selected.

As a rule, no interactions are to be expected.

Selection of packaging materials for semi-solid dosage forms

Suitable tests have to be carried out.
Packaging: Aluminium tube internally lacquered, plastic tubes.
Problems: Corrosion of metal tube; interaction with internal lacquering; sorption, permeation of water vapour, oxygen, aromas, essential oils.
Tests packaging material – dosage form: To test for corrosion, the filled metal tubes are stored horizontally, upright and inverted at 40 °C for 3 months and are then investigated.

The test for permeation and sorption, the filled plastic tubes are stored for 1 month at 40 °C and for 3 months at 30 °C/70% r.H. (At 40 °C the tubes can easily tear open at the end, which is why the storage period is restricted to 1 month.)

When selecting the packaging material, the climatic zone in which the product is to be introduced must also be taken into account.

Because of the problems arising with plastic tubes, aluminium tubes are preferred.

Selection of packaging materials for liquid dosage forms

Packaging: ampoule, injection vial with rubber stopper, glass bottle or plastic bottle with screw closure or pilferproof closure and liner.

Problems: pH, leakage, desorption, sorption, permeation, interaction with rubber stopper, interaction with liner.
Tests packaging material – dosage form: To test for sorption, permeation, pH and leakage, the finished product solution is filled in the container, and for desorption placebo solution is used. The samples are stored vertically and inverted under the following conditions: 30 °C/70%, 40 °C, 50 °C for up to 12 weeks
Testing intervals: 0, 1, 2, 3 months

4. Long-Term Tests

Step 4 is the central part of the stability tests.

These tests accompany scaling-up, i. e. development from the laboratory to the production scale with validated batches.

The samples are stored under conditions directly equivalent to the various climatic conditions, and thus binding stability statements can be derived from the results. These form an important part of licensing or registration documents.

Their scope depends on whether appropriate stability information is already available from pilot and stress tests, in which case the stability tests will be only of a more formal nature, or on whether the necessary information still has to be generated. This also determines to a decisive degree the point at which a definite stability assessment of the product will be possible.

4.1 Objective

- ☐ Confirmation of the results of stress tests.
- ☐ Establishment of binding stability information for clinical samples.
- ☐ Determination of the influence of batch size on the stability.
- ☐ Derivation of binding expiration dates for
  - a country
  - a climatic zone
  - all climatic zones, worldwide
- ☐ Determination of use life, overage, if necessary.

4.2 Requirement Profile for Practical Implementation

The factors that potentially influence the stability can be classified as follows:

Influencing factors from manufacture to use: these include statements on the influence of the storage period, and about sensitivity to temperature, moisture, light and oxygen.

These factors can influence the stability during
- ☐ manufacture
- ☐ storage at the manufacturer
- ☐ transport
- ☐ storage at the wholesaler
- ☐ storage at the officinal or hospital pharmacy
- ☐ storage in the household medicine cupboard.

Influencing factors during product use:
- ☐ repeated opening of multiple dose containers
- ☐ preparations modified into an appropriate form immediately before use, e.g. powder for reconstitution

4.3 Schedule of Tests

☐ **Selection of batches, storage samples**

The laboratory, pilot plant and production batches
- must be representative of the manufacture and packaging,
- should if possible be produced by a validated manufacturing process at laboratory, pilot plant or production level,
- must comply in all test criteria with the tolerances for freshly manufactured product,
- the information required by the authorities must be provided. Manufacturer, batch size, manufacturer and batch code for active ingredient.

The stored samples must be representative of the batches.

The analysis samples must be withdrawn and processed in such a way that the result is representative of the batch at the time the analysis is performed.

☐ **Test criteria**

In stability testing the criteria of a product must be followed up and investigated
- which may undergo changes during the course of storage,
- which are particularly important for the quality, efficacy and acceptance.

The test criteria of quality control and of stability testing may therefore be different.

☐ **Analytical methods**

The analytical methods must ensure for all test parameters that any changes that may occur are identified conclusively and at the earliest possible stage.
- For organoleptic analytical procedures colour cards and shade scales may be used.
- For physico-chemical analytical procedures in which multiple determinations are possible without difficulty, statistics are used for the evaluation, e.g. hardness, disintegration time, dissolution rate.
- The analytical techniques for assay determination must be specific and validated, i.e. the following points must be taken into account:

- selectivity
- linearity
- accuracy/recovery
- precision
- limit of quantitation for decomposition products
- ruggedness

The necessary information about reference substances or standards should also be supplied at this point.

The apparatus for physical and physico-chemical procedures should be qualified.

Statements on the number of parallel determinations for one test sample must also be provided.

- Organoleptic tests: single determination
- Physico-chemical tests: variable, single determination e.g. for pH, particle size, viscosity multiple determination e.g. for disintegration 6, dissolution rate 6, hardness 20.
  - Assay determinations:
    RSD $\leq$ 1.5, initial analysis 3-fold, subsequent analysis usually single. If a result is outside the time and temperature grid, repetition as 3-fold analysis.
    RSD $\geq$ 1.5, generally 3-fold analysis

  Analysis means the entire determination from weighing out the test sample up to documentation.

☐ **Tolerances**

The tolerances apply up to the expiration date and include:

- range of variation of manufacture
- precision of the analytical technique
- tolerable changes during storage

The last point is especially important because it has a decisive influence on the expiration date. It is not easy to determine.

A stepwise approach is followed for this purpose:

- determination of the discriminatory power of the analytical technique
  - for assay determinations, generally by means of statistical data: $\bar{x}$, s, RSD
  - specifically batch-related for many physico-chemical test procedures by means of statistical data: $\bar{x}$, s, RSD
    On the basis of the statistical data, internal warning limits can be established that indicate the point from which a change becomes statistically significant.
- determination of tolerable changes
  - with validation samples
  - by means of stress tests

☐ **Storage conditions**

To allow stability testing of worldwide scope, the world has been divided into four climatic zones [5]. The following storage conditions were established for these zones: [3]

climatic zone	definition	storage condition
I	temperate climate	21 °C/45% r.H.
II	subtropical and Mediterranean climates	25 °C/60% r.H.
III	hot, dry climate	30 °C/35% r.H.
IV	hot, humid climate	30 °C/70% r.H.

These storage conditions relevant to the climatic zones are supplemented by two stress storage conditions:

4 °C, 40 °C/-, 40 °C/75%

Reasons:
Extensive systematic studies are not performed in all cases in development steps 2 and 3. With this approach, statements as to the possible influence of batch size during scaling-up can be made as quickly as possible. Stability predictions are also derived from the results of long-term tests by means of kinetic calculations.

The following possible storage conditions are obtained:

4 °C / –
21 °C / 45% r.H.
25 °C / 60% r.H.
30 °C / 35% r.H.
30 °C / 70% r.H.
40 °C / 75% r.H.

Which and how many of these are used depends on where and how far the drug product is to be distributed.

The following limitations can generally be applied:

- For solid dosage forms storage at 4 °C is ruled out.
- The storage condition 30 °C/35% r. H. is only necessary if a development is being pursued exclusively for countries of climatic zone III. Otherwise, this storage condition is covered by 30 °C/70%, 40 °C.

Examples of specific storage schemes are listed together with the frequency of tests and storage period.

☐ **Testing intervals and storage period**

The testing intervals depend on the stability and the development status of the drug product to be tested.

- unstable drug products, early development status:

0, 3, 6, 9, 12, 18, 24, 36, 48, 60 months

- stable drug products, production batches:
0, 6, 12, 24, 36, 48, 60 months

For the storage period, a distinction is made between stress condition and the storage condition for the climatic zones in which the product is to be distributed.

If a drug product is developed for climatic zone I, 25 °C/60% represents storage under stress conditions. If a drug product is developed for climatic zone II, 30 °C/70% represent storage under stress conditions.

From the storage conditions, testing intervals and storage period, the following schemes have been derived for worldwide use, with a distinction being made between solid and non-solid dosage forms, and between climatic zones I and II.

Climatic zone I

Solid dosage forms

temp. [°C]	rel. hum. [%]	storage period and testing intervals [months]									
		0	3	6	9	12	18	24	36	48	60
21	45	x	(x)	x	(x)	x	(x)	x	x	x	x
25	60		(x)	x	(x)	x					
40	(75)		x	x							

Semi-solid and liquid dosage forms

temp. [°C]	rel. hum. [%]	storage period and testing intervals [months]									
		0	3	6	9	12	18	24	36	48	60
4	—		(x)	x		x					
21	45	x	(x)	x	(x)	x	(x)	x	x	x	x
25	60		(x)	x	(x)	x					
40	(75)		x	x							

Climatic zone II

Solid dosage forms

temp. [°C]	rel. hum. [%]	storage period and testing intervals [months]									
		0	3	6	9	12	18	24	36	48	60
25	60	x	(x)	x	(x)	x	(x)	x	x	x	x
30	70		(x)	x	(x)	x					
40	(75)		x	x							

Semi-solid and liquid dosage forms

temp. [°C]	rel. hum. [%]	storage period and testing intervals [months]									
		0	3	6	9	12	18	24	36	48	60
4	—		(x)	x		x					
25	60	x	(x)	x	(x)	x	(x)	x	x	x	x
30	70		(x)	x	(x)	x					
40	(75)		x	x							

Worldwide

Solid dosage forms

temp. [°C]	rel. hum. [%]	storage period and testing intervals [months]									
		0	3	6	9	12	18	24	36	48	60
25	60	x	(x)	x	(x)	x	(x)	x	x	x	x
30	70		(x)	x	(x)	x	(x)	x	x	x	x
40	(75)		x	x							

Semi-solid and liquid dosage forms

temp. [°C]	rel. hum. [%]	storage period and testing intervals [months]									
		0	3	6	9	12	18	24	36	48	60
4	—		(x)	x		x					
25	60	x	(x)	x	(x)	x	(x)	x	x	x	x
30	70		x	x	(x)	x	(x)	x	x	x	x
40	(75)		x	x							

climatic zone	storage condition	stress condition
I	21 °C/45%	4 °C, 25 °C/60%, 40 °C
II	25 °C/60%	4 °C, 30 °C/70%, 40 °C
worldwide	25 °/60%, 30 °C/70%	4 °C 40 °C

The schemes for climatic zone I, II and worldwide include always stress conditions besides the corresponding storage condition.

It must be kept in mind, that the derived storage conditions for all climatic zones include a safety margin for temperature and partial pressure. Seasonal and daily variations in temperature are taken into account in the respective storage temperature.

The reasons to include also stress conditions are:

- to derive predictions in an early stage of development
- to find out as quickly as possible if the batch size has an influence on the stability
- to find out all possible physico-chemical changes
- to cover in the United States the parts of climatic zone IV.

□ **Number of batches**

From the viewpoint of stability testing, the development phase of a drug product extends from the first pilot tests until production becomes routine. The end of the first year of production can be regarded as the point of reference for this definition.

The following batches are necessary to reach a definite assessment of the stability behaviour:

- New active ingredient in new dosage form:
  3 batches
  3 from scaling-up.
- Subsequent development, in which available results can be consulted:
  2 batches
  1 from scaling-up, 1 from first year of production.
- Clinical trial in progress in this stage 4:
  1 batch
  period of testing: 2 years.

□ **Packaging materials**

A drug product is frequently marketed in several packaging materials which may differ considerably both in type and size. However, they do not all have to be included in a stability programme. This applies particularly for solid dosage forms.

If information about the sorption behaviour is available from the stress tests, conclusions can be drawn through inference by analogy. For example, if a permeable packaging material has been tested and is suitable, all packaging materials that are less permeable are suitable, including those of different sizes or with different closure systems.

Consequently, even a major change in a less permeable packaging material does not automatically necessitate new stability tests.

Solid dosage forms

The following packaging materials are used:

- blisters: PVC, polypropylene, composite film

- plastic tubes: polyproplylene, polyethylene
- glass bottles
- aluminium foil

Which of these is used depends on the stability of the drug product, the climatic zone for distribution and the market strategy.

If the drug products are known not to be extremely sensitive to moisture on the basis of the results of an acceleration test, the packaging materials can be used as follows in the individual climatic zones:

packaging material	climatic zones			
	I	II	III	IV
blisters	x	x	x	
plastic tubes	x	x	x	x
glass bottles	x	x	x	x
aluminium foil	x	x	x	x

Semi-solid dosage forms

Frequently used packaging materials:
- aluminium tube, internally lacquered
- plastic tubes
- plastic containers

General suitability in the four climatic zones:

packaging material	climatic zones			
	I	II	III	IV
plastic tubes	x	x		
plastic containers	x	x		
aluminium tubes, internally lacquered	x	x	x	x

For the packaging materials for semi-solid dosage forms it may be necessary to include different sizes or a major change in the stability programme.

Liquid dosage forms

Frequently used packaging materials:
- glass bottles with closure
- ampoules
- ampoules, glass bottles with rubber stopper
- plastic containers

General suitability in the four climatic zones:

packaging material	climatic zones I	II	III	IV
plastic containers	x	x		
glass bottles	x	x	x	x
ampoules	x	x	x	x

Different sizes are of no importance for ampoules or glass bottles.
A major change in plastic containers or elastomers must always be clarified by means of stability tests.

☐ **Evaluation**

The analytical results are evaluated in several steps:

- Batch per storage condition:
  For each test parameter it is investigated whether a significant change has occurred compared to the initial value. It is then established whether the tolerance limit was reached. The laws of statistics are applied for the evaluation.
- The storage conditions of a batch:
  The results obtained in the different storage conditions are compared.
- Different packaging materials of a batch:
  If different packaging materials are put into storage, the results are again compared.
- Different batches:
  The last stage is the critical comparison of individual batches.

In all these cases statistics are employed as an aid to evaluation.

The following information is then derived from the results:

- For which test parameter was the tolerance limit reached or exceeded? What where the influencing factors? Storage condition, packaging material?
- Were any differences seen between the packaging materials?
- Did the batches show the same stability behaviour in scaling-up?

All the results obtained during the course of development are then used to formulate the general information:

- active ingredient stability profile
- results of stress tests
- results obtained with analogous formulations
- results obtained with other dosage forms containing the same active ingredient

The stability information is then derived from all the data. If the expiration date has not yet been covered by the full storage period, a stability prediction must also be performed. For this purpose, a stability prediction based on the laws of reaction kinetics [6, 7] is performed for the active ingredient for the required climatic zones on the basis of the results of the stress tests and the long-term tests.

It is then established whether any other test parameter has failed to reach the expiration date predicted for the active ingredient assay.

The stability information covers all test criteria, and the parameter which is or will be the first to reach the tolerance limit is the one which has a limiting effect on the expiration date.

Finally, it is decided whether it is necessary to ensure compliance with the expiration date which has been established by providing storage instructions to be displayed on the pack.

Storage period up to registration

If 6 months values of one and 12 months value of two representative batches are available, stability information about the product as a whole can be derived. The predicted expiration date is between 2 and 3 years, depending upon the stability behaviour and the available results.

The long-term tests are continued until the expiration date and storage period correspond.

☐ **Stability information**

All the results and the stability information derived therefrom are compiled in a stability report. This contains:

- the batch information,
- the stability testing specifications,
- the results,
- a critical assessment of the analytical techniques used and of the results,
- a statement by which the expiration date is limited and as to whether it may possibly be extended,
- the expiration dates for the individual climatic zones,
- storage instructions
- use life
- overage

} if necessary

Storage instructions

Storage instructions to be displayed on packs must reflect a genuine necessity, must take into account the prevailing temperatures of the respective climatic zone and must be derived directly from tests.

A logical relationship to the storage conditions thereby results.

When establishing the storage instructions, a very careful examination is necessary to identify the factors which influence the stability. The following generalising statements can be made:

- **Store at up to 30 °C**

  This will relate mainly to physico-chemical instabilities. If the temperature is exceeded, there will be an irreversible change in the short or medium term. The information about this is either known, e.g. suppositories, or is obtained on storage at 40 °C.

- **Store at up to 25 °C**
  In this case too, the instabilities will be of a physico-chemical nature. The samples stored at 30 °C exhibit this change in the medium or short term, but at 25 °C they are stable.
- **Store at 2–8 °C, refrigerate, do not freeze**
  These are chemical instabilities (sensitive active ingredients).
- **Store below 8 °C**
  These are chemical instabilities.
- **Store below – 18 °C**, deep freeze
  Sensitive active ingredients.

Use life

The same use life is generally applicable for eye drops. For all other relevant drug products it must be determined experimentally. In some cases, storage instructions on the packs will be necessary.

Overage

If an overage is necessary, the precise reason must be stated. An overage is not included in order to compensate inadequate pharmaceutical-technological optimisation of the formulation.

5. Follow-Up Tests

Step 5 comprises the investigations which are performed after licensing or registration, as long as no major change occurs.

5.1 Objective

☐ On-going stability testing
- Monitoring of stability for as long as the expiration date and storage period are not yet identical.
- Stability testing on 3 production batches if these are not yet contained in licensing documents.

☐ Follow-up stability testing
- Monitoring of on-going production and transition to routine, as long as no major change occurs.
- Confirmation of the stability information derived in the development phase.

5.2 Schedule of Tests

The on-going stability testing is a continuation of long-term testing in all respects.

Follow-up stability testing

Selection of batches and samples, test criteria, analytical techniques, specifications, storage conditions must be identical with those of long-term testing, because only in this way can the stability information be confirmed and any deviations recognised.

□ Storage conditions

25 °C/60% r. H.

40 °C/(75% r. H.)

- Testing intervals and storage period

temp. [°C]	rel. hum. [%]	storage period and testing intervals [months]						
25	60	0		12	24	36	48	60
40	(75)		6					

- Number of batches

  1 per year.

  If several dosages are on the market concurrently, only the lowest is tested. If several packaging materials are on the market concurrently, testing is performed only with the most sensitive.

6. Investigations in the Event of Changes During Continuous Production, Major Change

Step 6 comprises the stability tests that become necessary in the event of a major change. The major change may be a

□ change in the composition

□ change in the manufacturing process

□ change in the packaging material

It is frequently difficult to distinguish between a minor and major change.

6.1 Objective

- Investigation whether a major change adversely influences the derived stability information.

- The result must be available quickly to ensure the continuity of production.

6.2 Schedule of Tests

- Selection of batches and samples
  Frequently an experimental batch is the first one available. Parallel to this, a production batch of the unchanged dosage form.
- Test criteria, analytical techniques, specifications as for long-term testing.
- Storage conditions
  25 °C/60%
  30 °C/70%
  40 °C/(75%)

□ Testing intervals, storage period

temp. [°C]	rel. hum. [%]	storage period and testing intervals [months]								
25	60	0	3	6	9	12	24	36	48	60
30	70		3	6	9	12				
40	(75)		3	6						

□ Number of batches: 2

□ Evaluation
After 6 months an investigation is performed to determine whether the stability behaviour of the two dosage forms is identical.
- If no differences are observed, the stability information to date remains valid. Further examinations are made during follow-up stability testing.
- If differences are found, the expiration date must be shortened, and if necessary stress and acceleration tests must be performed.
  Number of batches: 2

7. Summary

A general concept for stability testing was presented which covers all the stages of development of a drug product extending from the first pilot test to continuous production including a major change. The overall development is divided into six steps, and each step takes into account the 12 principles of stability testing. Once again, these are:

□ systematic development

- ☐ selection of batches and samples
- ☐ test criteria
- ☐ analytical methods
- ☐ tolerances
- ☐ storage conditions
- ☐ testing intervals
- ☐ storage period
- ☐ number of batches
- ☐ packaging materials
- ☐ evaluation
- ☐ statements

The requirements derived from the principles are in some cases fixed from the outset, e.g. selection of batches and samples, storage conditions, testing intervals, and in some cases they also pass through a process of development, e.g. test criteria, analytical techniques, tolerances, evaluation, statements.

The principle of systematic development however ensures that the individual steps are logically coordinated with each other, even if practical experience shows that a development does not always proceed as logically as it is often described afterwards.

In this way it is ensured that the knowledge relating to the drug product is continuously improved since the test results are logically interrelated and can be consulted for each subsequent stage of development or in order to produce the overall statements.

The general concept is distinguished by a high degree of flexibility:

- ☐ A development can be entered at any stage and existing results can be built upon. In this way it is possible to establish a stability programme that is individually tailored to each problem but which still fits into the general concept.
- ☐ During a development, effort can continue to be directed specifically according to whether the drug product is to be introduced in a country, a climatic zone or worldwide.

The general concept was prepared taking into account the most important requirements of the regulatory agencies relating to stability testing.

Since stability testing accompanies the development of a drug product in a stepwise manner, it is possible to derive the stability information stepwise, but also to extend it stepwise.

Thus it is also possible to provide the stability information requested by individual authorities while a drug product is still being developed.

CTX, IND, Clinical Phase II and III Documents

For the licensing documents necessary at the end, stability information is available which incorporates the results of the entire development and not only 6- or 12-month values for two or three batches. In conclusion, it can therefore be stated:

The general concept for stability testing presented is

- ☐ comprehensive,
- ☐ flexible,
- ☐ problem-orientated,
- ☐ rational,
- ☐ draws fully upon the experience and skill of the developer,
- ☐ can be applied to new dosage forms or developments at all times.

The stability information obtained ensures that the perfect quality of a drug product can be guaranteed up to the expiration date.

This stability information is thus also suitable for registration documents to be submitted in the different countries.

At the same time, the general concept represents the basis for the desired harmonisation of the guideline for stability testing.

References

[1] Grimm, W., APV Symposium Stability Testing, Darmstadt 1991

[2] PMA's Joint-PDS Stability Committee "Stability Concepts" Pharmaceutical Technology, June 1984

[3] Grimm, W., Drugs Made in Germany **29**, 1986

[4] Grimm, W., Unpublished Tests 1991

[5] Futscher, N., Schumacher, P., Pharm. Ind. **34**, 479–483 (1972)

[6] Grimm, W., Schepky, G., Stabilitätsprüfung in der Pharmazie, Theorie und Praxis. Editio Cantor Verlag, Aulendorf 1980

14 Cyclodextrins: structure, application, stability and their influence on the stability of active substances

L. Szente, H-Budapest

Many new and to a greater extend old, generic drugs require new formulations to a higher standard than it was considered satisfactory several years ago. The search for new formulations is undoubtedly less expensive and results in a faster appearance of a product on the market, than to find and register a new substance.

In the field of drug formulations molecular encapsulation of active substances with cyclodextrins represents a rather novel way and seems to play a more and more decisive role.

This formulation way allows even the design of more effective drug delivery systems without any chemical modification of the active substance.

The first market success of the beta-cyclodextrin formulated Piroxicam (Brexin and Cicladol) by Chiesi and Masterpharma has emphasized the actuality and practical importance of the molecular encapsulation technique. The most recent price of beta-cyclodextrin enables its use even in the formulation of less expensive drugs, while several years ago it did not seem to be feasible due to the rather high price of the beta-cyclodextrin.

The non-toxic 2-hydroxypropyl-beta-cyclodextrin in parenteral dosage forms proves to be one of the most remarkable solubilizer for a great number of poorly soluble drugs. In the coming years several long known drugs are expected to be put on the market in the formulation with cyclodextrins and cyclodextrin derivatives.

1. Structure and types of Cyclodextrins of pharmaceutical significance

The cyclodextrins are produced by enzymic degradation of starch thus they belong to the group of enzyme-modified starch derivatives. Three different parent cyclodextrins are known: the alpha-cyclodextrin which consists of 6 glucose units, the beta-CD (7 glucose units) and gamma-CD (8 glucose units) (Fig. 1). All of them are crystalline, chemically uniform, non-hygroscopic substances. The doughnut-shaped cyclic molecules have an internal cavity of well defined dimensions. These molecular cavities are rather hydrophobic, while the outer surface of the cyclodextrin molecules are of hydrophilic character.

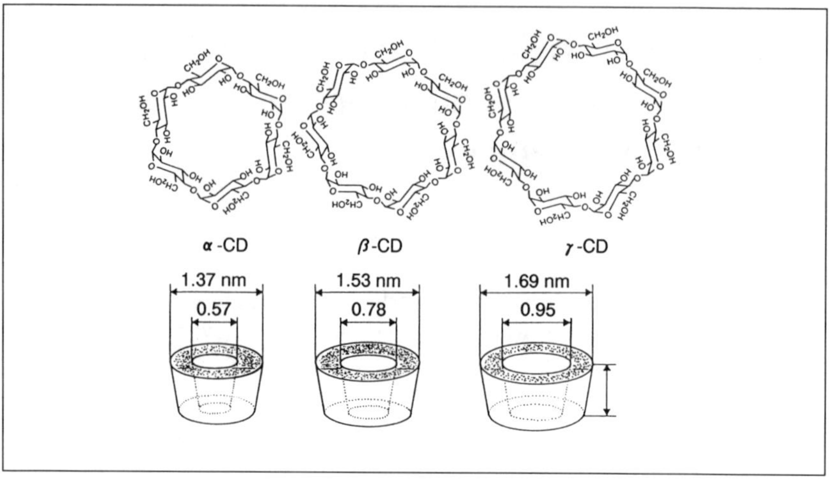

Fig. 1: Structure of cyclodextrins and their geometrical dimensions

Molecules or certain functional groups of molecules that are less hydrophilic than water can be included into the cavity of a cyclodextrin molecule in the presence of water. In an aqueous solution the slightly apolar cyclodextrin cavity is occupied by water molecules that are energetically unfavoured and these water molecules can be replaced by an appropriate guest substance present. This substitution process can be considered as essence of the molecular entrapment (encapsulation on molecular level) of the potential guest molecule (Fig. 2). The formed generally biner adduct is an inclusion complex, in which the guest substance is reversibly bound by apolar-apolar interaction and in part by hydrogen bonds but never by any chemical interaction. The formed inclusion complexes readily dissociate in presence of water, particularly in presence of a third component (a potential competitive guest substance). The phar-

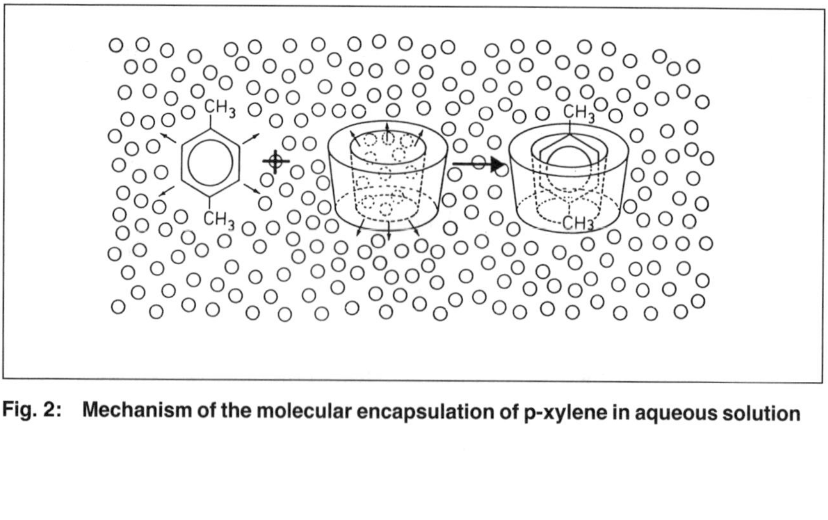

Fig. 2: Mechanism of the molecular encapsulation of p-xylene in aqueous solution

maceutical utilization of cyclodextrins also involves the use of chemically modified cyclodextrins (especially substituted beta-cyclodextrins) usually of higher water solubility.

Practically four types of cyclodextrin derivatives are at present applied:

- alkylated cyclodextrins (methylated and ethylated ones)
- hydroxyalkylated cyclodextrins (hydroxypropylated and hydroxyethylated ones)
- branched cyclodextrins (enzyme produced glycosyl- and maltosyl derivatives)

2. The utilization of cyclodextrins and their derivatives in pharmaceuticals

The utilization of the molecular encapsulation of biologically active substances by cyclodextrins usually aims at:

- modification of physicochemical properties of drugs
- improvement of stability in physical and chemical sense
- enhancement of bioavailability of poorly soluble and absorbable drugs
- to ensure content uniformity in solid dosage forms at extremely low doses
- reduction of undesired side effects of drugs [1].

All of these improvements are achieved without any chemical modification of the active substance which is of practical importance in terms of registration and approval of new formulations.

3. Primary consequences of the molecular encapsulation of drugs by cyclodextrins

3.1 Phase transformation, changes of the crystallinity

Liquid, oily or waxy substances can be transformed by cyclodextrin complexation into an easy to handle, stable, microcrystalline, non-hygroscopic powder. The pharmacologically active, natural hive product, propolis is a multicomponent paste-like material which is difficult to formulate into any solid dosage form. The beta-cyclodextrin complexation of propolis results in a microcrystalline, stable product of well defined crystallinity, [2] as illustrated by the X-ray powder diffractogram of the complex (Fig. 3).

The same is true for the inclusion complexes of natural essential oils, plant extracts, oleoresins, etc. All these oily extracts exhibit in complex form rather high crystallinity and can be characterized well by X-ray diffractometry. Even such drugs which have well defined crystal structure after cyclodextrin comple-

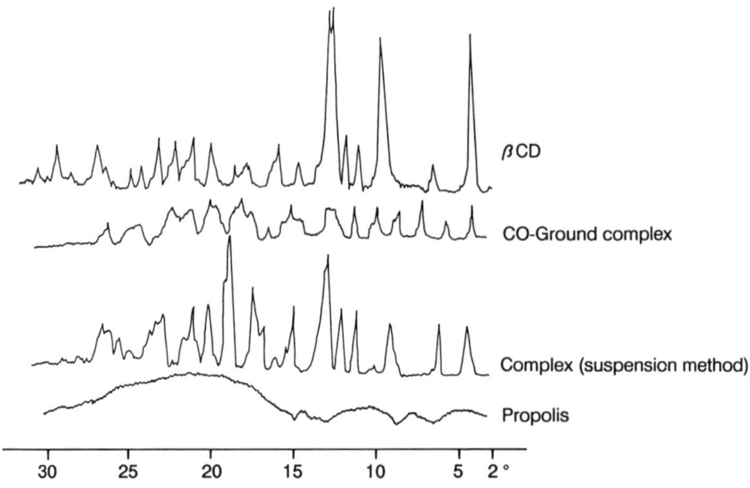

Fig. 3: X-ray diffraction pattern of natural propolis, the crystalline beta-cyclodextrin hydrate and the propolis-beta-cyclodextrin complexes

xation appear to possess new crystalline characteristics, due to the molecular encapsulation. In some cases complexation – in particular with cyclodextrin derivatives – may result in a product of amorphous character [4].

3.2 Solubilization by cyclodextrin complexation

Systematic studies have been performed in the past 15 years on the solubilization of poorly water soluble drugs by using parent and chemically modified cyclodextrins. It has been found that the water solubility of all drugs studied increases with increasing cyclodextrin concentration (Fig. 4), [3]. The extent of the solubilization depends on the stability constant of the complex in solution under given conditions. It has been observed that the solubilizing power of the chemically modified – highly soluble beta-cyclodextrin derivatives – in most cases greatly surpasses that of the parent beta-cyclodextrin as shown on the example of tolnaftate in Fig. 4.a [3].

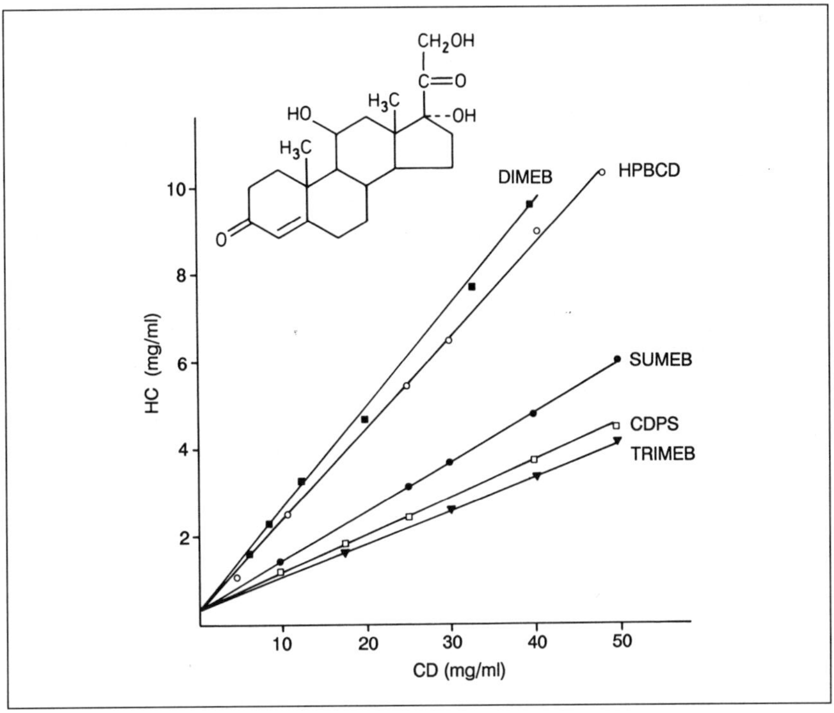

Fig. 4: Phase solubility diagrams of hydrocortisone in the presence of cyclodextrins

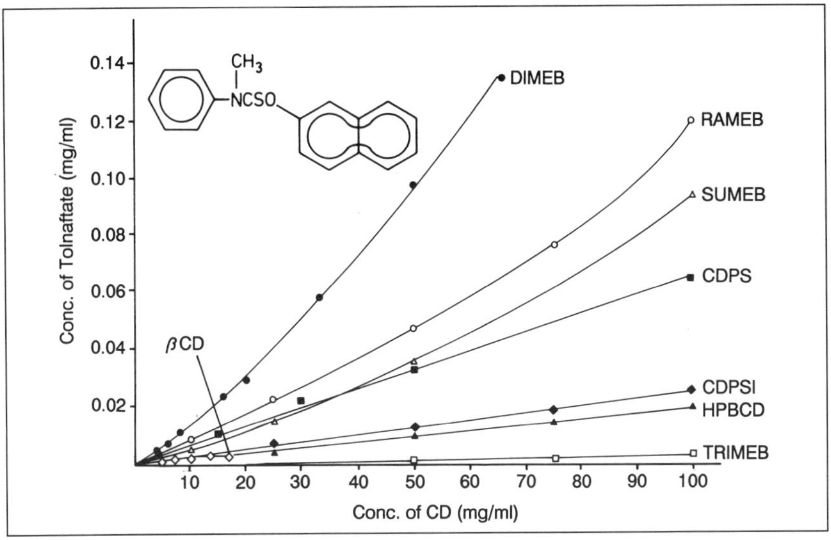

Fig. 4.a: Solubility isotherms of Tolnaftate in presence of parent and substituted beta-cyclodextrins

3.3 Enhancement of the rate of dissolution

Lipophilic drug substances in molecularly encapsulated form are in a preformulation which in general posses the following characteristics:

- the active substance is in a molecularly dispersed state (solid solution)
- the wettability of the solid form is greatly improved in water
- the dissolution rate of the entrapped drug is enhanced

The above advantageous properties are in particular true for the inclusion complexes of drugs with highly water soluble cyclodextrin derivatives. Due to the molecularly dispersed state of active substances and the increased hydrophilicity the dissolution process is greatly accelerated under physiological conditions [4]. In the presence of a third component (a potential guest molecule) even a programmed-release formulation can be designed. On the example of steroids (triamcinolon-acetonide, hydrocortison) is illustrated how molecular encapsulation improves the dissolution rate of poorly soluble steroids in aqueous media under normal conditions (Fig. 5) [4].

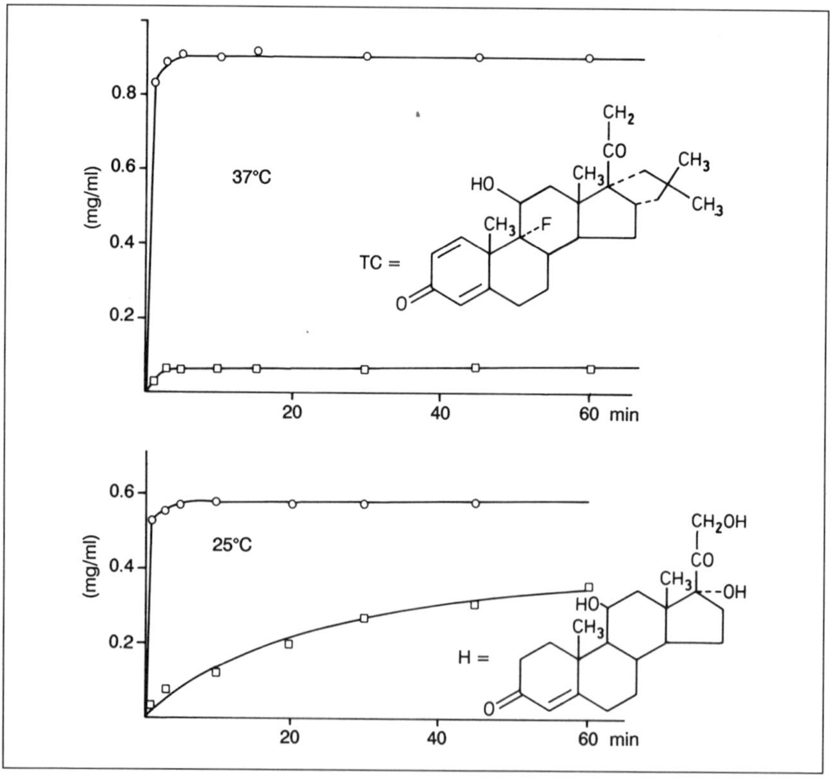

Fig. 5: Rates of dissolution of free and beta-cyclodextrin complexed triamcinolon-acetonide and hydrocortison in water

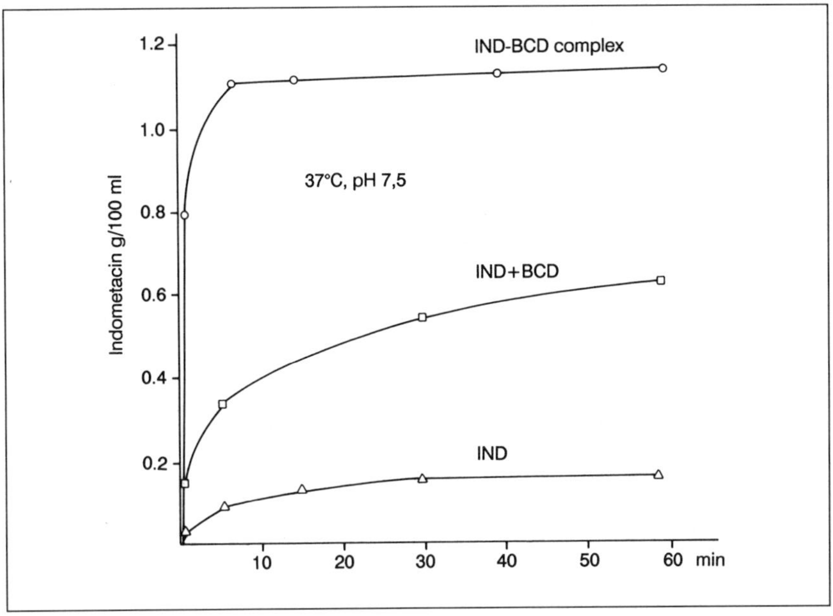

Fig. 6.a: Dissolution profile of free, beta-cyclodextrin complexed indomethacin and their mechanical mixture

The enhanced rates of dissolution moreover may lead to an improved biological activity of the complexed drugs. The acid form of the non-steroidal antiinflammatory agent, indomethacin when simply mixed or complexed with beta-cyclodextrin shows a remarkably increased dissolution profile compared to that of the free drug [5], (Fig. 6.a). As seen on Fig. 6.b the *per os* administered indomethacin-beta-cyclodextrin formulation resulted in a significantly enhanced blood level of the drug in rats [5].

The nasal dosage form of insulin complexed with 2,6-di-O-methylated-beta-cyclodextrin was reported to provide a greatly increased serum insulin level in rats, in comparison with the traditional form of the hormone [6]. As Fig. 7 shows not only the actual insulin level of the blood increased significantly, but also an improved biological response (reduced serum glucose level) was observed. This is a direct proof that cyclodextrin complexation in most cases results both in enhanced drug blood level and improved biological effectiveness, thus arises the possibility of the eventual reduction of the dose of a given drug when complexed with cyclodextrins. New type of transdermal [7], sublingual [8], and ocular [9] controlled release dosage forms have been designed utilizing 2,6-di-O-methyl-beta-cyclodextrin or 2-hydroxypropylated-beta-cyclodextrin as solubilizers. At present at least five different drugs are expected to have a successful market position in complexed form with highly soluble cyclodextrin derivatives of low toxicity.

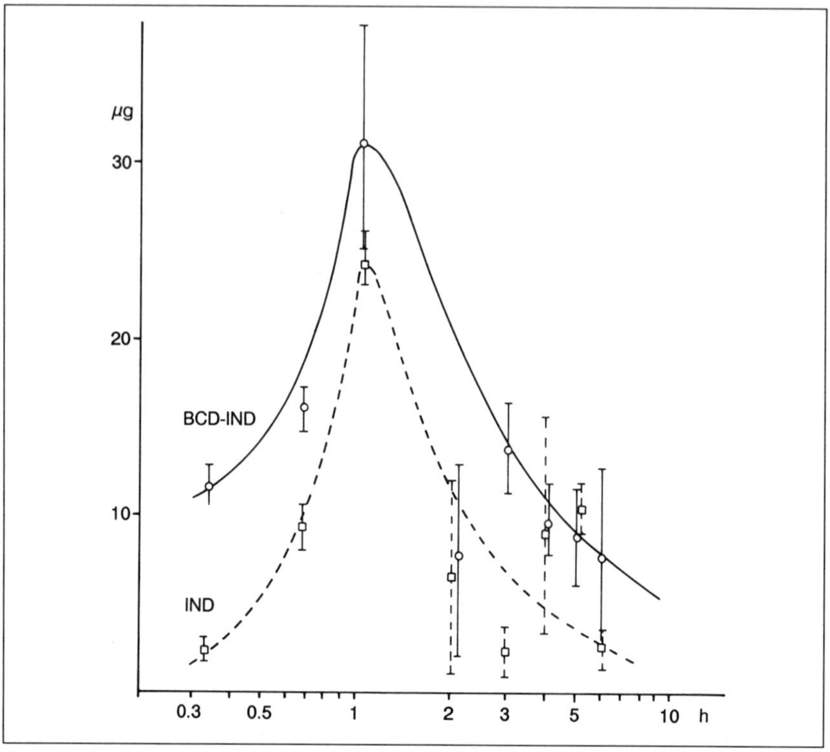

Fig. 6.b: Blood level of free and complexed indomethacin after *per os* application in rats

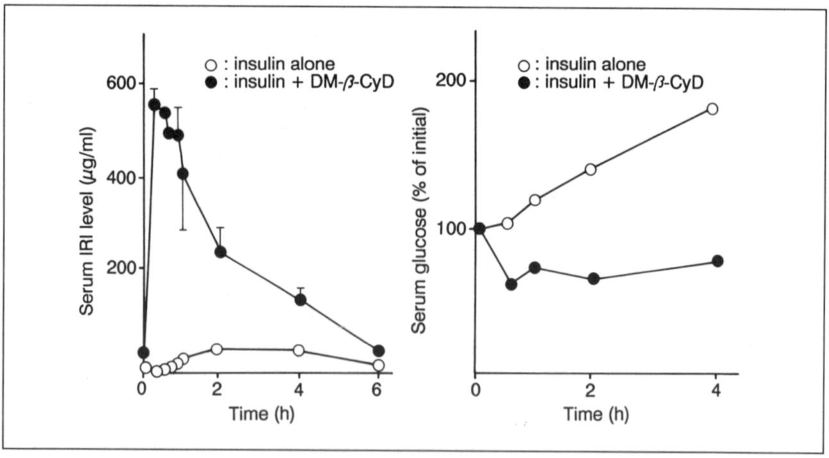

Fig. 7: Serum levels of insulin and glucose after nasal administration of free and 2,6,di-O-methyl-beta-cyclodextrin complexed insulin

3.4 Stabilizing effect of the molecular encapsulation with cyclodextrins

Encapsulation processes usually aim at the improvement of the stability of sensitive active substances against surroundings [10]. The cyclodextrin complexation is actually an encapsulation method carried out on molecular level, at which every single molecule of the guest substance is entrapped by the host cyclodextrin. This unique feature of the cyclodextrin complexation thus eliminates any kind of intermolecular changes between a multicomponent system, when formulated with cyclodextrins. The stabilizing effect of cyclodextrin complexation appears both in solid state and in solution, however, these effects are more pronounced in solid state, than in solution where the dissociation of the complex gradually takes place, due to the presence of water. (Inclusion complexation is a reversible process).

3.4.1 Stabilization against heat

Certain drugs are volatile substances which have considerable vapour pressure even at ambient temperature. Every naturally occuring essentials oils, plant extracts of pharmaceutical importance need to be stabilized against loss of consituents on heating, processing or during storage. More, than 40 different essential oils, oleoresins and plant extracts have successfully been stabilized against evaporation with beta-cyclodextrin complexation [10]. On the example of the antibacterial natural pine oil and its beta-cyclodextrin complex is illustrated that upon a ten-year storage of the complex under normal conditions in solid state not only the total essential oil content cut also its chemical composition remains practically unaltered (Fig. 8).

The volatility of L-menthol is remarkably reduced via inclusion complexation with beta-cyclodextrin. Thermal Evolution Analysis (TEA-Assay) undoubtedly has proved that the thermal release of the volatile substance takes only place above 100 °C from the beta-cyclodextrin complex, while the free or adsorbed L-menthol escapes entirely from the system by 100 °C (Fig. 9) [11].

Accelerated heat stability test-storage of the formulations at elevated temperatures for longer period of time provide quick informations on the stabilizing effects of complexations. Vitamin D_3 has been known to form a quite stable inclusion complex with beta-cyclodextrin in a molar ratio of 1:2. This formulation appears to resist to a storage at 60 °C for days without significant losses of the cholecalciferol (Fig. 10) [12].

Similar heat stability was reported for the beta-cyclodextrin complexed vitamin K_3 (menadione) even in the presence of an aminoacid mixture, where the heat sensitivity of the vitamin is more pronounced [13], (Fig. 11).

Polyene antibiotics show high sensitivity toward heat, but the molecular encapsulation with a proper-size cyclodextrin (cyclooctaamylose) a significant heat resistance is ensured. Fig. 12 shows that Flavofungine, a potent antibiotics is successfully protected against heat in gamma-cyclodextrin complexed form [14].

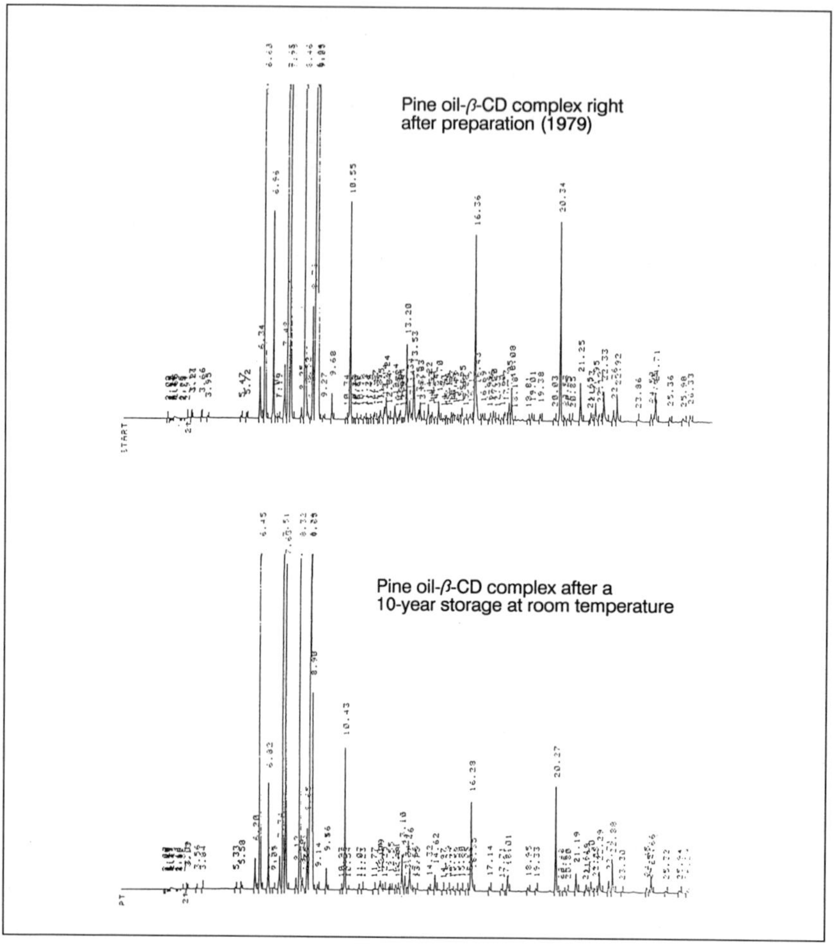

Fig. 8: Capillary gas chromatograms of freshly prepared pine oil-beta-cyclodextrin complex and that after a ten-year storage under normal conditions

It is important to note, that accelerated heat stability tests in certain cases might result in unexpected and disappointing results. It means that eg. the complexed substance decomposes faster and to higher extent that non-complexed species itself. Such observation may point to the possibility that a solid phase inclusion catalysis takes place upon heating [16]. This is especially true for solid phase transacylations, where the ester-type guest loses its acyl group attached later onto the cyclodextrin hydroxyls. It has been reported that aspirin when complexed with heptakis-2,6-di-O-methyl-beta-cyclodextrin decomposes faster than the non-complexed drug, particularly under higher humidity [15].

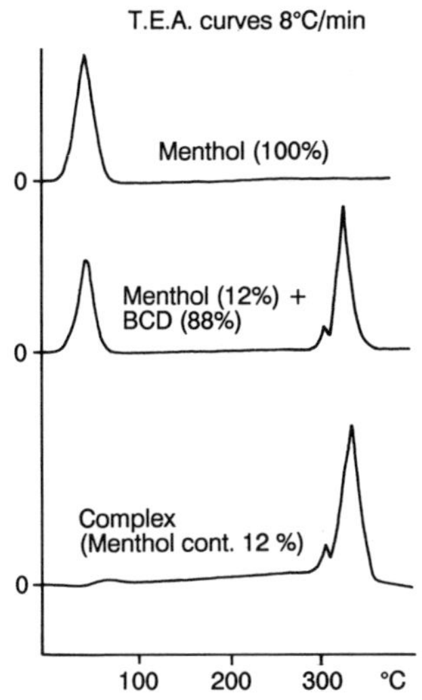

Fig. 9: TEA-analysis of L-menthol-beta-CD mixture and the inclusion complex

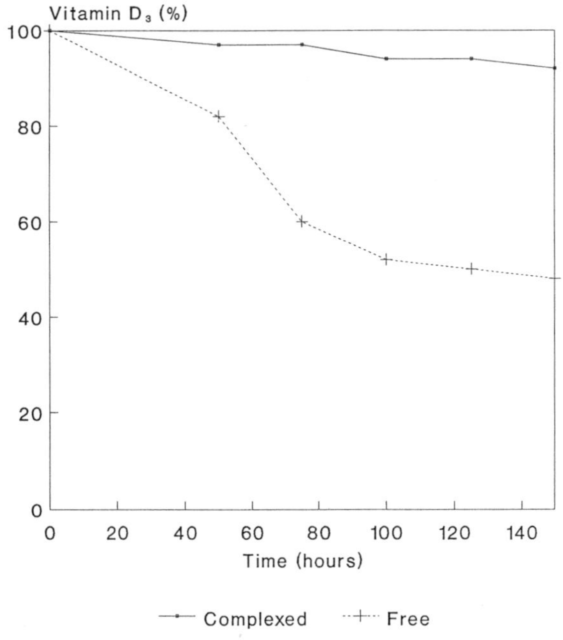

Fig. 10: Thermostability of free and complexed vitamin D_3 at 60 °C in solid state

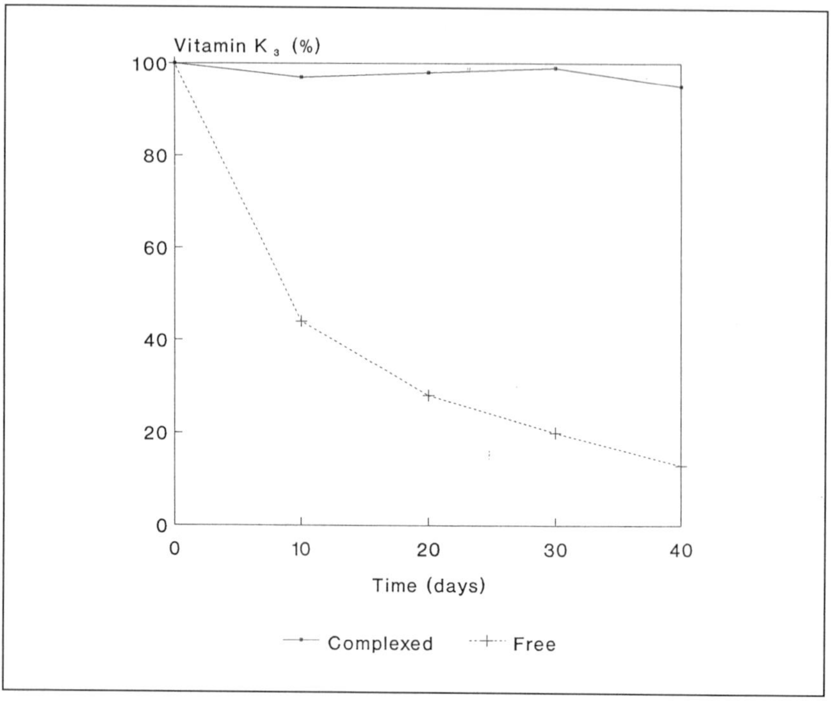

Fig. 11: Thermostability of free and beta-CD complexed vitamin K_3 at 60 °C in presence of an aminoacid mixture, in solid state

These results suggest that accelerated stability tests of solid cyclodextrin complexes are to be carried out only at 40–60–70 °C, but never above 80 °C, if the guest substance is suspected to generate such acyltransfer processes.

3.4.2 Stabilization against oxidation

Molecular encapsulation of the easily oxidizable drug substances provides protection against oxygen. The successful stabilization of the oxygen-sensitive guest molecule requires a special mode of inclusion into the cyclodextrin cavity where the oxygen susceptible part of the drug is hidden by cavity wall. Linolic and linolenic acid in beta-cyclodextrin complex of 1:4 resp. 1:2 molar ratio have been found to be protected against oxydation even stored in pure oxygen atmosphere for two weeks [10]. This result indicates that probably the oxygen sensitive double bonds are somehow surrounded by the cyclodextrin ring. The highly oxygen sensitive benzaldehyde and cinnamaldehyde have been transformed into a oxygen-resistant formulation by beta-cyclodextrin complexation. These complexes have been stored in a modified Warburg-apparatus in pure

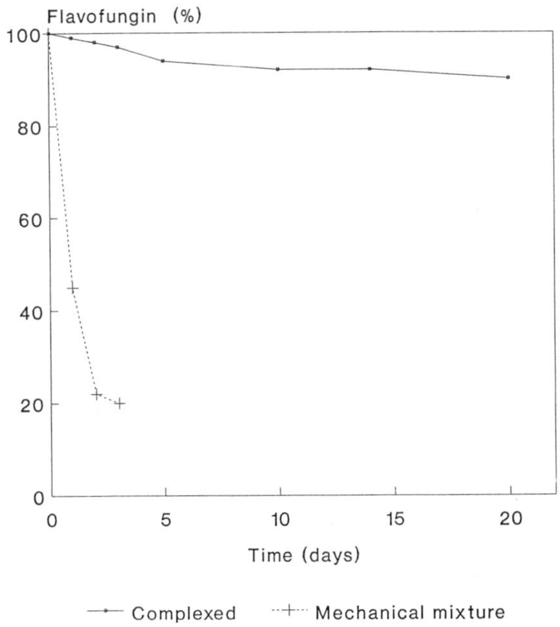

Fig. 12: Accelerated stability test of the adsorbed and gamma-cyclodextrin complexed flavofungine at 60 °C in solid state

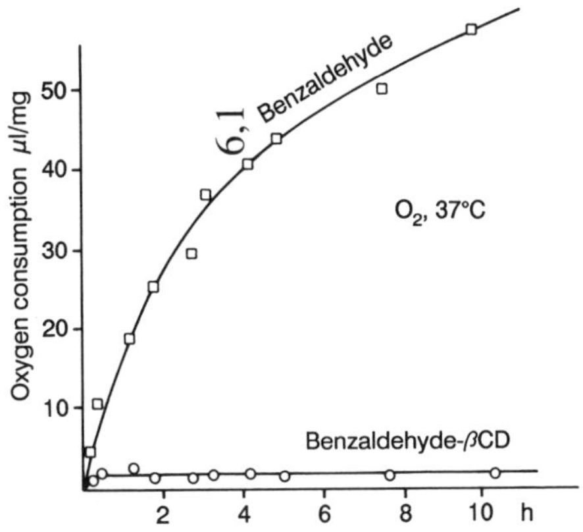

Fig. 13: Oxygen consumption of free and complexed benzaldehyde by Warburg method

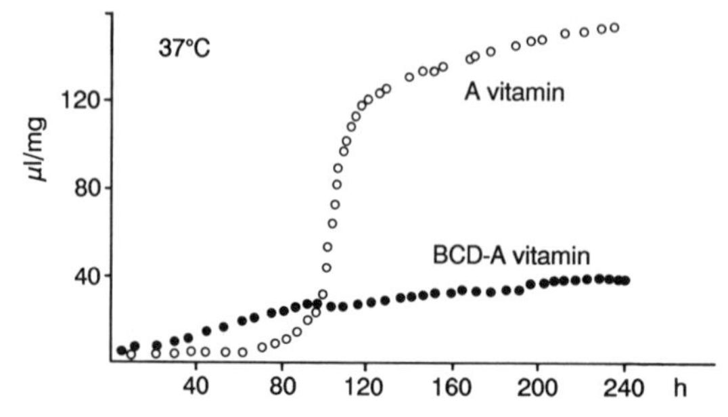

Fig. 14: Oxygen uptake of free and beta-CD complexed vitamin A, determined by Warburg technique

oxygen atmosphere for days and the oxygen consumption of the free and complexed benzaldehyde was continuously measured (Fig. 13).

Similar oxidation kinetics was registered on the free and beta-cyclodextrin included vitamin A with Warburg-technique, as shown in Fig. 14 [12].

3.4.3 *Stabilization against light induced decompositions*

The photodecomposition of several drugs can also be reduced by molecular entrapment with cyclodextrins and cyclodextrin derivatives.

The photodegradation of antitumor benzaldehyde has considerably been retarded due to the formulation with either alpha-, beta- or gamma-cyclodextrin. The most powerful stabilizing eectit was obtained with alpha-cyclodextrin complexation, while beta- and gamma-cyclodextrin had less pronounced protection effect, as seen in Table 1.

Under certain light stress conditions, however, specific solid state transformations may occur, which similarly to the heat stress-treatments, leads to the

Table 1: Photodegradation of benzaldehyde and its cyclodextrin complexes under 2500 lux, at 30 °C, in solid state [17]

sample	remaining intact benzaldehyde (%) at time (min)				
	15	30	60	90	120
free benzaldehyde	44	31	12	7	7
alpha-CD complex	97	94	92	90	90
beta-CD complex	90	88	79	77	74
gamma-CD complex	87	83	67	64	60

possibilities of unusual photochemical reactions due to the special mode of inclusion of the guest substances in the cyclodextrin cavity. In this case again, cyclodextrin cavity acts as a molecular reaction vessel with a given hydrophobic environment. It has been observed that the solid beta-cyclodextrin-benzaldehyde complex when treated under UV light (313 nm) for 3 hours will be transformed into an optically active benzoin with an enantiomeric excess of 15%. This photodecomposition product is normally observed only when benzaldehyde is irradiated in benzene solution [18]. Thus the stress light stability tests may be misinformative and must be interpreted with care.

The photo-decomposition rates of free-beta- and gamma-CD complexed clofibrate revealed to the stabilizing effect of complexation against irradiation [19] as illustrated in Fig. 15.

Cholecalciferol complexed with beta-cyclodextrin in a molar ratio of 1:2 showed excellent stability against photodecomposition induced by UV-irradiation [12]. As shown in Fig. 16 non-complexed vitamin D_3 decomposes significantly faster than the molecularly entrapped form.

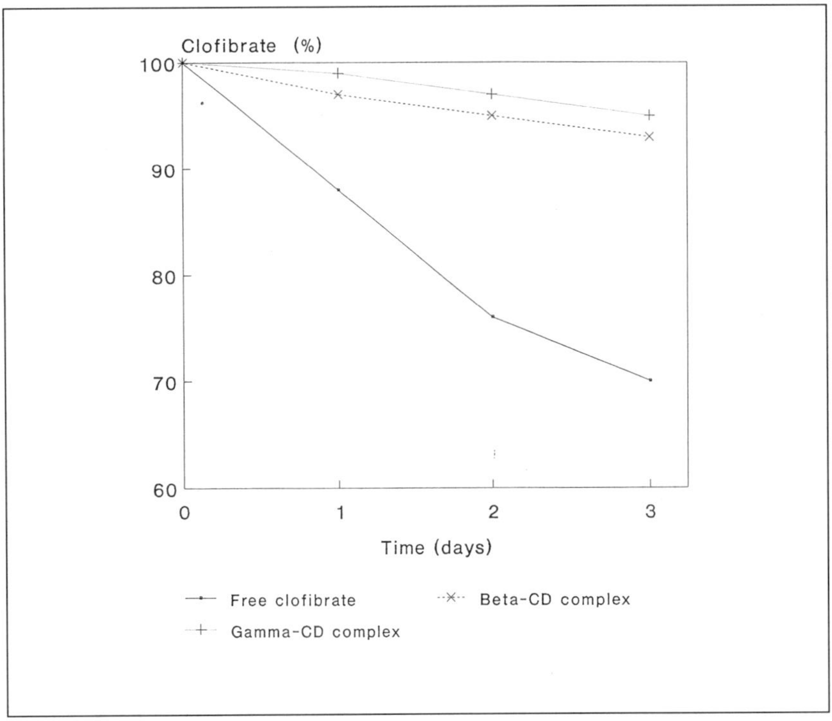

Fig. 15: Photodecomposition of free and complexed clofibrate at 30 °C in solid state

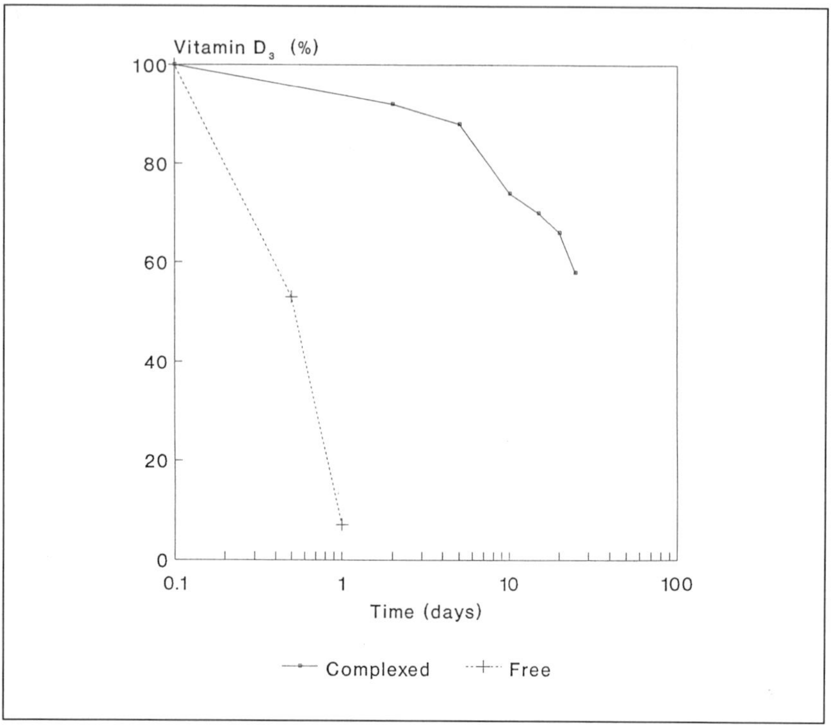

Fig. 16: Light stability of free and beta-cyclodextrin complexed vitamin D_3 in solid state

3.4.4 Stabilizations in aqueous solution

The stabilizing effect of the cyclodextrin complexation appears most remarkably in solid state, however, in most cases the increase of stability occurs even in aqueous solutions when the dissociation of the inclusion complex gradually takes place, continuously generating free, non-complexed species. It has moreover been observed that even the presence of cyclodextrins causes stability enhancement, without previous complex formation, in situ complex formation takes place.

Proscillaridin showed to be significantly stabilized in the presence of both beta- and gamma-cyclodextrin at pH 1,46, while the presence of alpha-cyclodextrin did not retard the hydrolytic decomposition of the drug [20]. Similar tendencies were observed regarding the hydrolytic degradation of Thalidomid at pH 8,5. In this case again alpha-cyclodextrin was unable to protect Thalidomid against decomposition, but beta- and gamma-cyclodextrin showed stabilizing effect [21], (Fig. 17). In the both above cases alpha-cyclodextrin cavity proved to be just too small for the bulky guests, thus the inclusion of those drugs could not take place.

Among prostaglandins many derivatives are known to be sensitive against hydrolytic processes resulting in either isomerization or degradation which lead to the alteration or loss of their biological activity. Prostaglandin I_2-methyl ester has been transformed into an inclusion complex with beta-cyclodextrin and this complexed form was found to be stable against hydrolysis in aqueous solutions [22]. The hydrolytic degradation kinetics of free and complexed prostaglandin I_2-methyl ester at different pH was studied in details and the results are shown in Fig. 18.

The chemically modified cyclodextrins sometimes provide even higher stabilizing effect than parent cyclodextrins (which have 18, 21, or 24 free hydroxyl groups respectively) probably due to the difference in number and location of hydroxyls, and equilibrium water content. The water sensitive guest eg. can be stabilized to a much higher extent with heptakis 2,3,6-tri-O-methyl-beta-cyclodextrin, than with parent beta-cyclodextrin, since the total absence of rim hydroxyls excludes any inclusion catalytic (alkoxi-effect) phenomena and in addition the solid complex shows no moisture sensitivity, at all.

Hydroxyethylated – beta-cyclodextrin was reported to reduce the hydrolytic degradation of indomethacin in aqueous solution at pH 6,6 upon a six-month storage [23].

Heptakis-2,6-di-O-methyl-beta-cyclodextrin was the only effective one among studied cyclodextrins in protection of the antitumor drug (SKF 104 864-H) against lactone-ring hydrolysis [24].

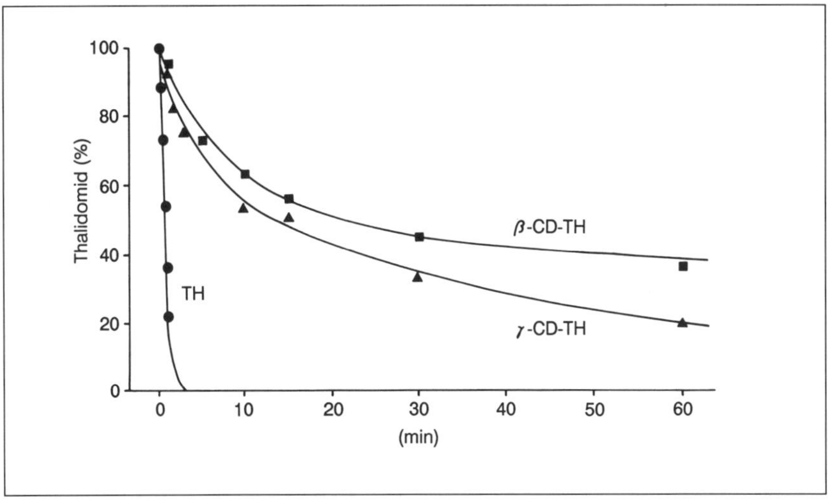

Fig. 17: Hydrolytic decomposition of Thalidomid in absence and presence of cyclodextrins at pH 8,5

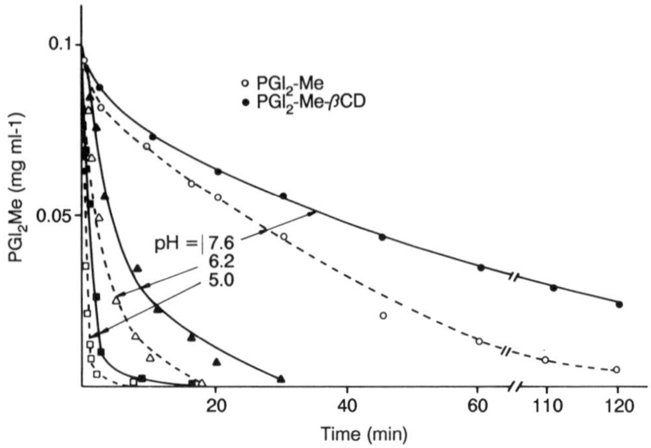

Fig. 18: Degradation kinetics of prostaglandin I_2-methylester and its beta-cyclodextrin complex at different pH

The photodecomposition of chlorpromazine at pH 4,0 in aqueous solution was successfully decelerated by cyclodextrins both under aerobic and anaerobic conditions [25].

Emetine and cephaline were found to be protected against thermal and photochemical degradation in solution by using gamma-cyclodextrin and heptakis-2,6-di-O-methyl-beta-cyclodextrin as host substances [26].

One of the most frequently used chemically modified cyclodextrin, the 2-hydroxypropyl-beta-cyclodextrin was reported as a apotent solubilizer and stabilizer for chlorambucil in aqueous solution at different pH [27].

Many of the biologically active peptides and proteins require stabilization against chemical decompositions and aggregation upon storage. The majority of the injectable protein preparations are exposed to a possible denaturation and thus deactivation due to the minor changes in their three-dimensional structure. Hydroxypropylated-cyclodextrin was found to improve the stability of such proteins (insulin, interleukin-2) in buffered aqueous solutions [28].

In certain cases again, the effect of cyclodextrin complexation may result in obvious acceleration of the decomposition process of a given drug. (The inclusion catalytic phenomena was first observed in aqueous solutions on the example of different nitrophenyl-acetate-cyclodextrin systems [29].

This possible inclusion catalytic effect can be expected especially in case of ester type drugs, where the mode of inclusion of the guest substance in the cyclodextrin cavity results in a the proximation of the esterbond of drug and a cyclodextrin rim-hydroxyl. Therefore the results of stability studies sometimes are just the reverse of those expected. In such cases the eventual existence of an inclusion catalytic process has to be taken into account. A cyclodextrin accelerated degradation of beta-lactam antibiotics has been recently reported [30].

Heptakis-2,6-di-O-methyl-beta-cyclodextrin, 2-hydroxypropyl-beta- and 2-hydroxypropyl-gamma-cyclodextrins and even glucose were found to accelerate the lactam-decomposition of phenoxymethylpenicillin and aztreonam. Both aztreonam and the penicillin derivative followed a pseudo-first order kinetics on their degradation, and all studied cyclodextrins signficantly fastened the decomposition process. As the above examples show, the cyclodextrin complexation usually results in improved stability of included drugs, however, depending on the modes of inclusion the complexation may even accelerate the degradation of certain biologically active substances [31].

4. Market situation and approved drugs containing cyclodextrins

At present six different products are marketed which contain drug-cyclodextrin formulation. Five of those products contain beta-cyclodextrin and only one is formulated with alpha-cyclodextrin.

The composition, indication and manufacturers are listed in Table 2.

Table 2: Presently approved and marketed drugs formulated with cyclodextrin

Composition	Trade name	Dosage form	Indication	Manufactur
PGE_1-alpha-cyclodextrin	Prostavasin	intraarterial	vasodilator	ONO, Japan Schwarz, Germany
PGE_2-beta-cyclodextrin	Prostarmon	sublingual tablet	labour initiator	ONO, Japan
Piroxicam-beta-cyclodextrin	Brexin Cicladol	tablet suppository	analgetic and antiphlogistic	Chiesie and Masterpharma Italy
Benexate-beta-cyclodextrin	Lonmiel Ulgut	capsules	antiulcer	Shionogi, Japan Teikoku, Japan
Iodine-beta-cyclodextrin	Mena-Gargle	gargling	disinfectant	Kyushin, Japan
Garlic oil-beta-cyclodextrin	Xund Allidex	dragees	anti-atherosclerotic	Bipharm, Germany Pharmafontana, Hungary

References

[1] Szejtli, J., Pharm. Technol. Int. **3**, 3. March 16 (1991).

[2] Szente, L., J. Szejtli, Acta Pharm. Technol. **33**, 218 (1987).

[3] Szemán, J., L. Szente, T. Szabó, Proc. 4th Intl. Symposium on Cyclodextrins p. 164. (1984).

[4] Szejtli, J., Cyclodextrin Technology, Kluwer Publ. Co., Dordrecht, p. 186, 1988.

[5] Szejtli, J., Staerke **33**, 387 (1981).

[6] Arima, H. et al., Minutes, 5th Intl. Symp. on Cyclodextrins p. 487. de Santé, Paris (1989).

[7] Loftsson, T., N. Bodor, Acta Pharm. Nord. **1**, 85 (1989).

[8] Pitha, J., J. Pharm. Sci. **75**, 165 (1986).

[9] Usayapant, A., H. Karara, Pharm. Res. **6**, S91 (1989).

[10] Szente, L., Ph. D. Thesis Univ. Budapest, 1978.

[11] Szente, L., M. Gál, J. Szejtli, J. Incl. Phen. **2**, 631 (1984).

[12] Szejtli, J., E. Bolla, Staerke **32**, 386 (1980).

[13] Tardy, L., J. Szejtli, J. Incl. Phen. **3**, 1 (1985).

[14] Vikmon, M., Proc. 4th Intl. Symp. on Cyclodextrins, München, p. 307 (1984).

[15] El-Gendhy et al., Int. J. Pharm. **31**, 25 (1986).

[16] Wernick, D. et al., Abstr. of the 4th Intl. Symp. on Inclusion Phen. S2–S3, Lancaster 20–25 July (1986).

[17] Uekama, K. et al., Int. J. Pharm. **13**, 253 (1983).

[18] Rao, P., N. Turro, Tetrahedron Lett. **30**, 35 (1989).

[19] Uekama, K. et al., Pharm. Acta Helv. **58**, 338 (1983).

[20] Uekama, K. et al., Acta Pharm. Suec. **20**, 287 (1983).

[21] Koch, H., P. Steinacker, Arch. Pharm. **321**, 287 (1988).

[22] Szemán, J., J. Szejtli, Acta Pharm. Technol. **33**, 27 (1987).

[23] Brauns, U., Acta Pharm. Technol. **32**, 153 (1986).

[24] Underberg, W., J. Pharm. Biomed. Anal. **8**, 681 (1990).

[25] Uekama, K., Jpn. J. Antibiotics *32*, S-103 (1979).

[26] Teshima, D., Chem. Pharm. Bull. **37**, 6. 1591 (1989).

[27] Loftsson, T., Int. J. Pharm. **66**, 289 (1990).

[28] Brewster, M. et al., J. Parent. Sci. Technol. **43**, 231 (1989).

[29] Van Etten, C. et al., J. Am. Chem. Soc. **89**, 3242 (1967).

[30] Loftsson, T., Int. J. Pharm. **67**, R5–R7 (1991).

[31] Beyrich, T., J. Wagner, Pharmazie **44**, 448 (1989).

15 Stability and Stability Testing of Biopharmaceuticals

H. Hoffmann, S. Bassarab, M. Schlüter, W. Werz und R. G. Werner, D-Biberach/Riß

Introduction

Recombinant DNA-derived proteins and monoclonal antibodies are currently used in medicine for a broad spectrum of indications, such as endocrine disorders, cardiovascular diseases, cancer, viral diseases, organ transplantations, immunological disorders, antidotes and others. Besides therapeutic use of proteins, they are applied as vaccines, drug carriers and as diagnostic tools.

A large variety of proteins is currently being investigated for use as drugs or devices including hormones, enzymes, antibodies, antigens, immune factors and others. The access to these proteins has been made available by hybridoma- and recombinant DNA-technology, supported by the methods of modern biotechnology [1, 2]. These technologies made it possible to manufacture large amounts of complex protein drugs at high quality and purity which had been previously unobtainable or too expensive and difficult to produce, or could only be isolated from natural sources with a higher risk for product safety [3].

The appropriate production system for such protein drugs depends on the nature of the protein which is intended to be produced. Bacteria as host cells for recombinant DNA are suitable for non-glycosylated proteins with low molecular weight and a small number of disulfide bridges. Yeast cells and mammalian cell culture systems are more appropriate for the production of more complex glycoproteins with high molecular weight and numberous disulfide bridges [4].

The structure and function of proteins is determined by their amino acid sequence which defines the peptide backbone (primary structure). The folding or the shape of the backbone in a alphahelix, β-sheet or random coiled form (secondary structure), depends on the sequence and the interaction of amino acid residues (disulfid bridges, hydrogen-bonds, etc.). The 3-dimensional structure (tertiary structure) of a protein depends on the orientation of amino acid residues and their relationship to helical regions. In some cases individual protein subunits are non-covalently linked to larger protein complexes (quartiary structure).

The complexity and structure of proteins can also be influenced by posttranslational modifications, such as proteolytic processing, glycosylation, alkylation and deamidation. Two different types of glycosylation are distinguished: O-linked glycosylation with N-actetylgalactosamine-containing carbohydrate moieties linked to serine or threonine; N-linked glycosylation with N-acetylglucosamine-containing, large mannose- and complex-type carbohydrate moieties linked to asparagine. Glycosylation may influence the solubility of proteins, their biological activity and/or their pharmacokinetic behaviour. For many proteins the biological function of glycosylation is still unknown.

Proteins as drug substances differ from conventional chemical drugs by many properties such as size, shape and conformation, multiple functional groups, amphoteric nature, physical form of bulk drug and heterogeneity of structure. Therefore, proteins represent an order of magnitude greater complexity than traditional drugs. This complexity of protein drugs has an impact on the number of different analytical methods which have to be applied for protein characterization and for the description of stability indicating properties.

In spite of these differences, the current guidelines on stability and stability testing are also applicable to protein drugs [5, 6]. The requirements and the goals are the same: the drugs must be designed to withstand long periods of transport and storage. The full effectiveness and safety of industrially manufactured drugs must be guaranteed until the end of the declared shelf-life. This can only be achieved and ensured by an extensive stability program.

1. Modifications of the protein structure which may influence protein stability

An important property of proteins is their amphoteric nature, resulting from multiple ionizing groups. The pH value influences the net charge on proteins. The isoelectric point (pI) is characteristic for each protein and often corresponds to the pH of minimal solubility. This is due to the fact that at the isoelectric point the net charge of the protein equals zero.

Many purified protein preparations derived from natural sources or from biotechnical production consist of a population of differently charged molecules and therefore cover a pI-range instead of having an individual pI-point. This so-called microheterogeneity in the charge distribution is well-known for most of the glycoproteins and can be explained by a population of molecules which differ in the composition of monosaccharides, specifically sialic acids, in their complex carbohydrate moieties. The charge distribution of proteins can also be influenced by modifications of the chemical structure, particularly by deamidation of glutamine (Gln) and of asparagine (Asn). Therefore, the determination of charge distribution by isoelectric focussing is an important stability indicating assay.

Proteins are sensitive to several chemical modifications. Besides deamidation of Gln and Asn, oxidation of methionine (Met), beta elimination of cysteine (Cys), intra- and/or intermolecular disulfide rearrangement and non-covalent or covalent protein self-association can occur. Chemical modifications of proteins can be potentiated or induced by temperature, pH, light and composition of the formulation buffer. Such chemical modifications may result in alteration of structure, loss of function, acquired immunogenicity and altered pharmacokinetics.

Instability of protein preparations can also be caused by proteolytic degradation via proteases, which cleave peptide bonds within the peptide chain. The resistence of proteins to proteolytic cleavage depends on several main factors: purity, temperature, pH, concentration and accessibility of cleavage sites. Proteolytic digestion can be a problem during protein recovery, especially in early stages when the protein is in a less pure state. Proteases can be detected and characterized using synthetic peptide substrates which develop coloured reactions or by zymogram-techniques. Proteolytic degradation products of proteins can be detected by SDS polyacrylamide gel electrophoresis (SDS-PAGE) using sensitive silver staining techniques or by other suitable chromatographic techniques, like HPLC (high pressure liquid chromatography).

Protein denaturation is a major change from the original native structure without alteration of the amino acid sequence or cleavage of peptide bonds between the amino-acids. Factors that may induce protein denaturation include: elevated temperatures, pH-extremes, organic solvents, chaotropic salts, hydrophobic surfaces, air-liquid interfaces, interaction with metal ions and shear forces due to agitation. Protein denaturation may cause protein aggregation and precipitation, resulting in loss of function, acquired immunogenicity and altered pharmacokinetics.

Protein aggregation can be caused by covalent or non-covalent binding. Covalent aggregation occurs via disulfid bonds (aggregates separate under reducing conditions) or via free amino, carbonyl, carboxyl, hydroxyl etc. functional groups. Non-covalent aggregates are formed via hydrophobic or electrostatic interactions and often dissociate under denaturing conditions, such as treatment with 8 M urea solution or with sodium dodecyl sulfate. Soluble aggregates are measured by gelpermeation chromatography, non-soluble aggregrates are detected by laser-light-scattering and UV-absorption.

The retention of structural integrity of proteins is important for the maintenance of efficacy and safety of protein drugs. The retention of the protein structure is also important for the maintenance of key physicochemical properties, e.g. solubility and stability.

2. Extensive analytical characterization for biopharmaceuticals from biotechnical production

Quality control release and stability testing has to cover different features of proteins: size, identity, structure, sequence, concentration, purity, surface-charge, shape and activity. Each of these properties is described by several analytical methods, which in their entirety give the information about structural integrity and consistency, efficacy, purity and safety of biopharmaceuticals.

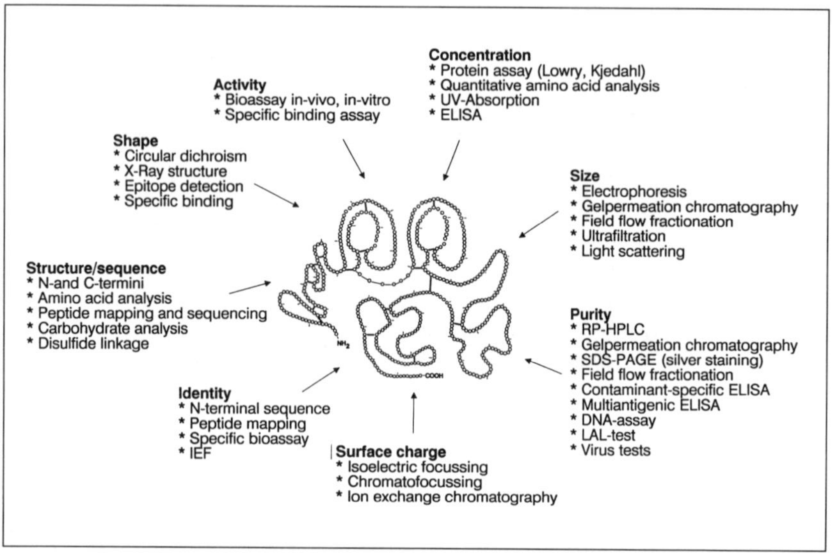

Fig. 1: Different features of proteins, described by a number of different test methods

---

▶

Fig. 2: Structure of human DNA-derived tissue plasminogen activator (t-PA)
T-PA consists of 527 amino acids (aa) with a molecular mass of 59050 without consideration of the carbohydrate moieties attached at positions aa 117, aa 184 and aa 448. The total molecular weight is in the range of 65000. The 3-dimensional structure of the t-PA is determined by the amino acid sequence and by 17 intramolecular disulfide bridges. The molecule is structured in 4 domaines: 1 fibronectin finger region (aa 1–43), 2. epidermal growth factor region (aa 44–91), 3. kringle 1 (aa 92–173) and kringle 2 regions (aa 180–261), 4. serine protease region (aa 276–527). The fibronection finger region and the kringle 2 region are involved in the fibrin-binding properties of t-PA, the protease region is involved in plasminogen activation. T-PA is cleaved by plasmin at position aa 275–276 into a two-chain form; A chain (kringle region) and B chain (protease region) are connected by a disulfide bridge. Both forms, single-chain and two-chain t-PA show comparable fibrinolytic activity in-vitro.

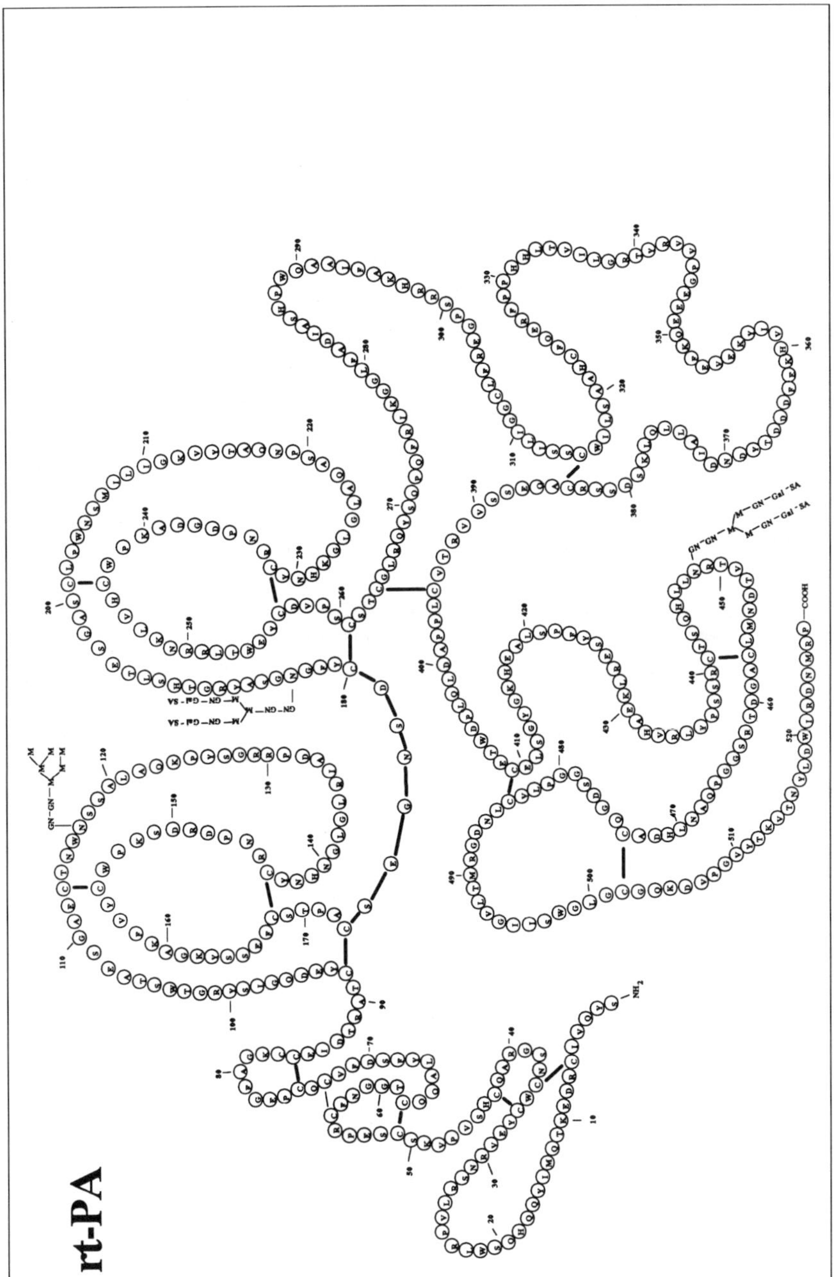
rt-PA

For each of these properties, a single assay is not sufficient to describe adequately the (unchanged) nature of the protein. Hence, several analytical methods are required. The variety of test methods used to describe and probe all these different features of biopharmaceuticals is given in Fig. 1.

Recombinant DNA-derived tissue plasminogen activator (t-PA) is a typical example of a complex glycoprotein [7]. The biological function of t-PA is the binding to fibrin clots and the activation of plasminogen, which leads to the generation of plasmin and to the degradation of fibrin clots or blood coagulates. T-PA is used as active ingredient in a pharmaceutical for fibrinolytic therapy in a number of different indications: myocardial infarction, lung embolism, deep veneous thrombosis and peripheral arterial occlusion [8, 9]. T-PA will be used as an example in the following to describe stability requirements and characteristics for biopharmaceuticals from biotechnical production.

The structure of t-PA is shown in Fig. 2. T-PA is a complex protein with 527 amino acids, divided in several domaines, containing 17 disulfide bridges and 3 glycosylation sites, with a molecular weight of 65000. For analytical release of each individual lot of Actilyse® about 30 different test methods are applied to prove efficacy, purity, safety and lot to lot consistency of this t-PA containing pharmaceutical.

In many of these tests a site by site comparison of the material to be released with a well-characterized reference standard is performed. A number of different analytical methods is applied for the characterization of reference standard. Some of these investigations are very extensive and time-consuming and will only be performed once for each product during development chemistry. These studies include the elucidation of the structure of the protein and of carbohydrate residues attached. As far as possible, the mechanism of action is investigated. During this development chemistry phase, analytical methods are developed and optimized which can later be applied on a routine base to detect deviations of a given protein structure. In many cases, the results of these methods reflect the heterogeneity of the protein, for example in the profile of a chromatogram or in the banding pattern of an electrophoretic gel.

Tryptic peptide mapping is an example of such a chromatographic method, which indirectly confirms the primary structure of the protein. T-PA is cleaved by trypsin after reduction and defolding into more than 50 peptids, which are separated by reversed-phase-HPLC to result in a characteristic fingerprint chromatogram. During development chemistry all of these peptides including glycopeptides have been isolated and identified by amino acid analysis and sequencing. The profile from tryptic peptide mapping is very sensitive to changes in the sequence or to chemical modifications of the peptides. Even single amino acid exchanges can be detected by comparison of the tryptic peptide map profile of the mutant protein with the profile of the reference standard. One peptide out of fifty shows an altered chromatographic behaviour due to an amino acid exchange in this peptide of the mutant protein (Fig. 3). Therefore, this sensitive technique indirectly confirms the primary sequence of

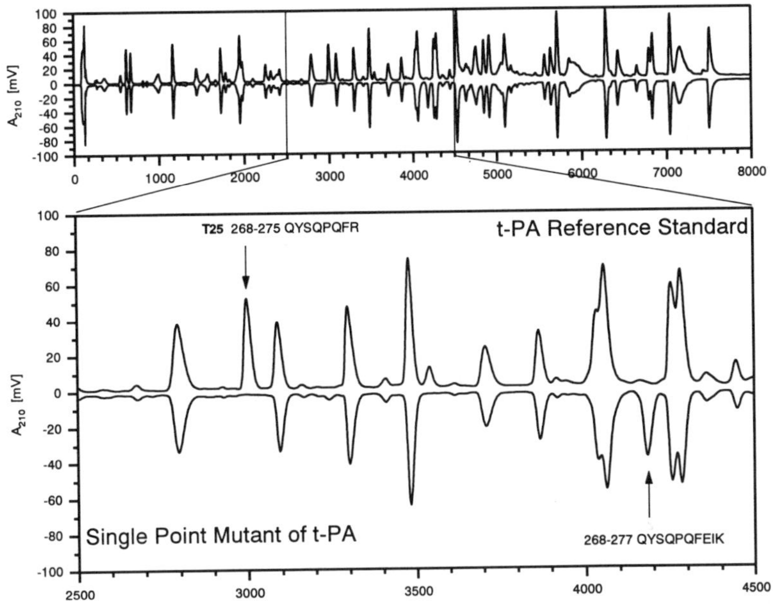

Fig. 3: Identification of a single point mutant of t-PA in comparison with t-PA reference standard by Tryptic Peptide Mapping. Exchange of Arg 275 for Gln 275

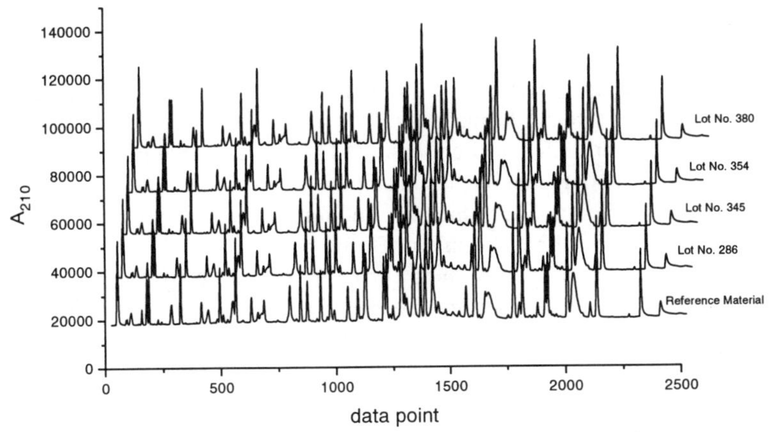

Fig. 4: Lot to lot consistency for different t-PA batches, demonstrated by Tryptic Peptide Mapping

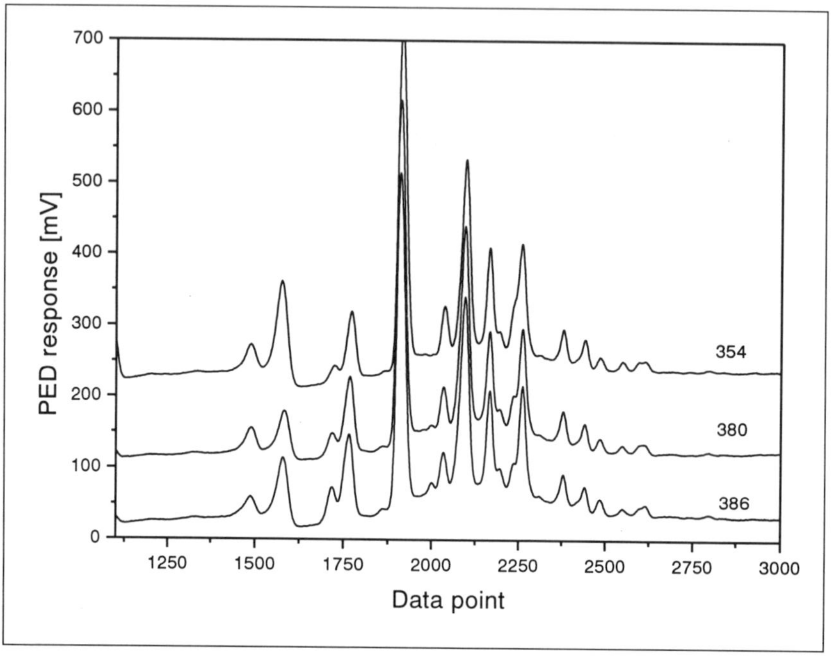

Fig. 5: Oligosaccharide Mapping for t-PA
Separation of neutral oligosaccharides by high pH anion exchange chromatography with pulsed amperometric detection (HPAE/PAD)

the protein drug and is routinely used for lot release testing. Lot to lot consistency can be demonstrated by comparative tryptic peptide mapping, as shown for different t-PA lots in Fig. 4.

The consistency in the carbohydrate structure of biopharmaceuticals can be analysed by oligosaccharide mapping. The carbohydrates are enzymatically cleaved from the protein and then separated by anion exchange chromatography at high pH and detected by pulsed amperometric detection. The profiles of the oligosaccharide maps are characteristic for the glycosylation pattern of a certain protein and reflect the microheterogeneity in the preparation. Fig. 5 shows a typical oligosaccharide map for t-PA, produced by chinese hamster ovary (CHO) cells under serum-free cell culture medium conditions. The microheterogeneity in glycosylation of proteins is also detected by isoelectric focussing, due to the different charge distribution in a population of slightly different glycosylated forms.

3. Process validation is the basis for lot to lot consistency and a key factor for a defined product shelf life

The basis for protein stability testing is a comprehensive product characterization with the emphasis on homogeneity of the preparation and lot to lot consistency. Pharmaceutical products are tested for pharmacokinetic parameters, efficacy and safety during clinical studies. It is therefore very important, that the product lots for the market are consistent with product lots in clinical phase III studies. Lot to lot consistency is also a key factor for a defined product shelf life of biopharmaceuticals.

Lot to lot consistency of biopharmaceuticals can be ensured by validated and well-controlled biotechnical production processes. The manufacturing of the active ingredient comprises a large number of different steps during cell culture, scale-up, fermentation and protein recovery. Due to the number of different process steps which may influence product quality and due to the complexicity of the product itself, the quality of biopharmaceuticals is always linked to the production process itself. Therefore, the manufacturing process has to be tightly controlled by defining standard operation procedures for all operations and by setting limits for all critical process parameters. A large number of inprocess controls has to be performed. All operations in the process have to be well documented. A process control system is often used in biotech plants for the automation and control of biotech processes, as well as for the documentation of process parameters.

Stability and solubility of the protein are a function of a number of physical and chemical conditions during processing. There are a number of critical steps which may have influence on the product quality: media conditions and the producing organism, fermentation technology, cell separation and harvest, first steps in protein recovery and purification, such as ultrafiltration, microfiltration and concentration steps, elution conditions during chromatography, final purification and final formulation. The compatibility and the interactions with manufacturing equipment and with container/closure surfaces have to be checked carefully. If the product is lyophilized, the conditions for the lyophilization process are also very critical.

Throughout the biotechnical and pharmaceutical manufacturing process of protein drugs great care has to be taken that there are no adverse effects on the product quality. The cells of the production organism are layed down in an extensively characterized cell bank. This defined cell bank is tested for stability beyond the production phase. The cells are grown under defined conditions in a fixed and validated fermentation process. After separation of the product from the production organism during harvest, the protein product is purified from other protein contaminants in the very first steps of protein recovery. Thereby, the product can already be transferred to a very pure and more stable form within a short period of time, avoiding proteolytic degradation of the product.

During final purification trace protein contaminants, potential viral contaminants and DNA are removed. Final formulation is the last step in the protein recovery scheme. The pharmaceutical manufacturing includes the filling in the final dosage form and lyophilization, if required for reasons of stability during shipment and storage.

During the development of the manufacturing process at pilot plant scale all factors which are known to influence product quality and product yield have been optimized. After scale-up to the commercial scale, 3–5 consistency runs are performed, which should prove the reliability and reproducibility of the process. An extensive analytical characterization of product from these consistency runs is performed. A consistent product quality has to be proven for material from these runs. The final product from the consistency runs is put on real time stability and the results from stability testing define the shelf-life of the product. The product from these consistency runs is used in clinical phase 3 studies and the first reference standard is established with material from these lots. The concept of lot to lot consistency is based on the goal of obtaining identical material of each production cycle, in order to ensure the same safety, efficacy and stability profile as for lots in clinical trials. This goal can be achieved by the establishment of a controlled and validated manufacturing process and comprehensive characterization of the protein drug during lot release testing. In addition, an extensive stability program is performed to establish the shelf life and guarantee the full effectiveness and safety throughout the declared shelf-life of a biopharmaceutical.

4. Definition of stability indicating methods

The results from analytical product characterization and first experiences concerning product stability during process development provide a good basis for the selection of stability indicating test methods. In many cases these tests are product-specific, such as potency or activity assays, or cover specific features of the individual proteins. On the other site, some general requirements for the scope of stability indicating test methods for biopharmaceuticals can be defined.

The stability indicating test methods should detect the most common degradation forms of biopharmaceuticals: inactive protein, soluble and non-soluble aggregates, proteolytically digested forms and chemical modified forms, such as deamidated protein. The stability indicating methods must be validated for specificity, linearity, accuracy, recovery rate, sensitivity, precision and limit of detection. The stability indicating property of each individual test method for a specific protein has to be proven.

Aggregate formation is one of the most abundant forms of protein instability. Non-soluble and soluble aggregates have to be distinguished. Soluble aggregates can be detected and quantified by gelfiltration chromatography. They do no necessarily lead to opalescence or turbidity of the protein solution.

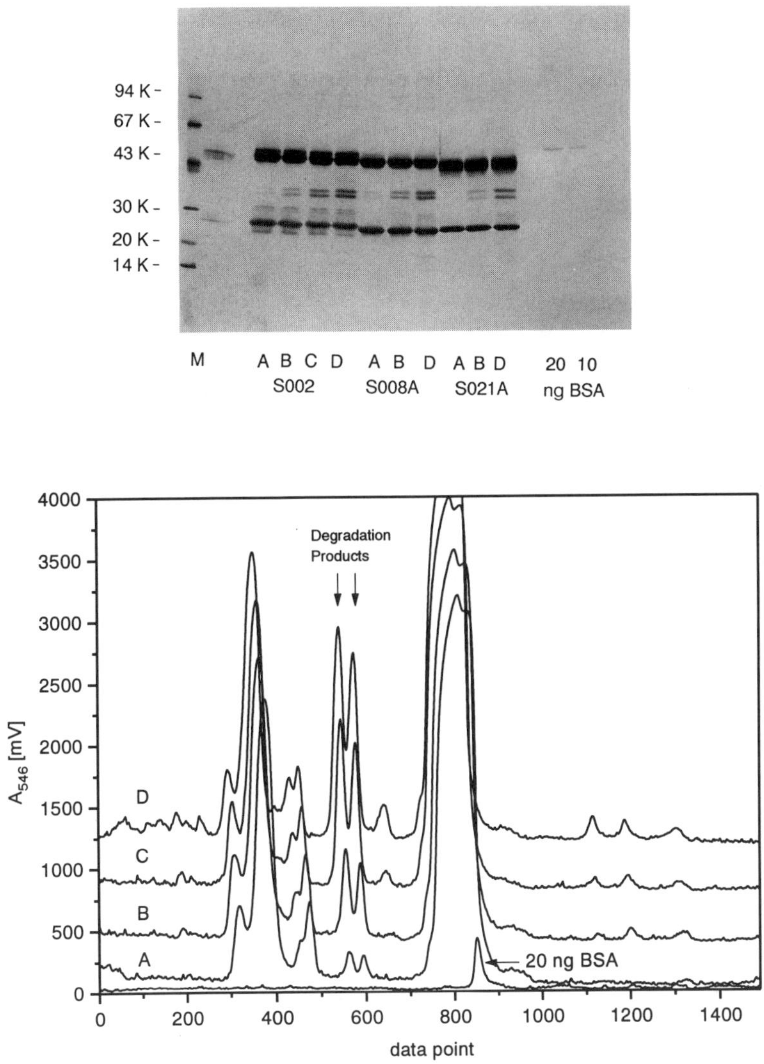

Fig. 6: Stability testing of a monoclonal antibodies using SDS-PAGE
Degradation products are visualized by SDS-PAGE (Silver Staining) (A) and quantified by densitometric scanning of the gel (B)
A: Reference Standard
B: Storage for 1 month at + 30 °C
C: Storage for 3 month at + 30 °C
D: Storage for 6 month at + 30 °C

In contrast, non-soluble aggregates consist of large protein particles which are not accessible to gelfiltration techniques. Already small amounts (< 1%) of non-soluble aggregates can cause turbidity in protein solutions. Non-soluble aggregates can be detected by light scattering techniques or UV-measurement. Field flow fractionation, a relatively new analytical technique, is capable to separate protein monomer from oligomeric forms and larger aggregates, including non-soluble aggregates within the same separation. This technique may prove to be very valuable to investigate the stability of proteins.

Protein degradation products can be visualized by SDS-PAGE (sodium dodecyl sulfate polyacrylamid gel electrophoresis) using a sensitive silver staining technique (Fig. 6A). The detection limit for proteolytic degradation products of proteins is in the range of 200–1000 ppm. Staining of SDS-gels with Coomassie-blue is ten times less sensitive, but more suitable for a quantitative measurement of degradation products by densitometric scanning of the gels. An example for the quantitative measurement of degradation products during stability testing of a monoclonal antibody by densitometric scanning of SDS-gels is shown in Fig. 6B.

Proteolytic processing of proteins can also be measured by chromatographic techniques like gelpermeation chromatography. T-PA has a predominant proteolytic cleavage site at position amino acid 275–276 and is readily cleaved by plasmin into two-chain t-PA. The ratio of single-chain to two-chain t-PA can be

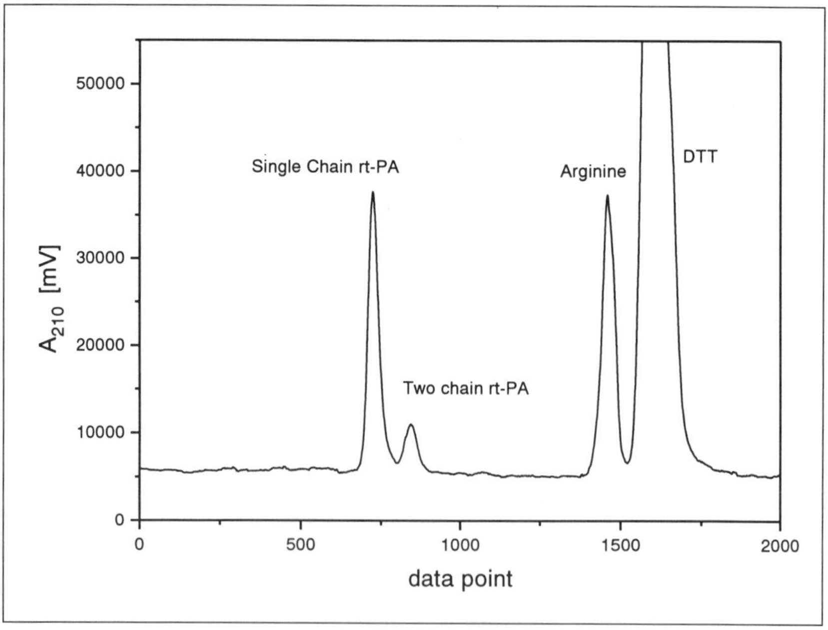

Fig. 7: HPLC-gelpermeation chromatography for the determination of the ratio of single-chain and two-chain t-PA

determined by HPLC-gelpermeation chromatography (Fig. 7). T-PA produced under serum-free conditions by mammalian cell culture exists predominantly in the single-chain form. During long term storage in liquid formulation this molecule will be cleaved into the two-chain form, depending on storage temperature and pH of the formulation buffer. Therefore, the HPLC-gelpermeation chromatography is a very important method for stability testing of t-PA.

As is true for all chemical pharmaceuticals, the measurement of potency of biopharmaceuticals is most important for quality control release as well as for stability testing.

5. Peculiarities for potency testing of biopharmaceuticals

For some protein containing pharmaceuticals only biological test systems (animal models or cell culture systems) are available. In-vivo animal models measure the "true" activity including pharmacokinetics of the protein but they normally suffer from a very high variability of test results. One example is the bioassay for insulin, which is measuring the blood glucose lowering ability in animals. Another test system is the measurement of weight gain in hypoxed rats after treatment with growth hormone. A potency assay for erythropoetin relies on the uptake of radiolabelled iron in mice.

For other proteins, especially for enzymes, good biomimetic in-vitro test systems have been developed. The reproducibility and accuracy is usually much higher than for in-vivo test models, but they strongly depend on the assay system itself. An example for an in-vitro test method with a relative broad range for assay variability is the CPE assay (cytopathic effect) for the measurement of antiviral activity of interferons. An example for a very precise and accurate in-vitro assay is the measurement of the fibrinolytic activity of t-PA in a clot lysis assay described below.

T-PA is a protease which cleaves plasminogen to plasmin an thereby initiates the lysis of fibrin clots and blood coagulates. Several in-vitro test have been developed for t-PA which measure the biological activity of t-PA (Fig. 8). The simplest test is the measurement of the protease activity of t-PA with the aid of a synthetic substrate, a tripeptide linked to p-nitroanilide (S-2288). This assay becomes more accurate and independent of the ratio of single-chain and two-chain form when plasmin is added, to transfer all t-PA present in the two-chain form, before addition of the substrate.

A higher specificity can be achieved in an indirect chromogenic assay where plasminogen is incubated with t-PA to create plasmin, which is measured using a synthetic, plasmin-specific substrate (S-2251). The amount of plasmin generated in this test correlates with the plasminogen-activating potency of t-PA.

The fibrinolytic activity of t-PA, which includes proteolytic as well as fibrin-binding properties of t-PA, is measured in an in-vitro clot lysis assay [11]. Four purified compounds are mixed at a defined time: fibrinogen, plasminogen, thrombin and t-PA. Three reactions occur: 1. Thrombin cleaves fibrinogen and

Direct chromogenic Assay

- T-PA + S2288 – cleavage of p-nitroanilide
  - measure absorbance at 405 nm
  - linear response to t-PA concentration
- Problem: different activity of 1-chain and 2-chaim forms
- Modification: t-PA + plasmin – 2-chain t-PA; measures total t-PA activity
- Utility:
  - Measurement of t-PA protease activity
  - Quantification of t-PA

Indirect Chromogenic Assay

- T-PA + plasminogen – plasmin
- Plasmin + S2251 – cleavage of p-nitroanilide
  - measure absorbance at 405 nm
  - linear response to t-PA concentration
- Utility:
- Identification of t-PA
  - Stability of t-PA
  - Quantification

In-vitro fibrinolytic Assay

- Fibrinogen + Plasminogen + Thrombin + t-PA
- 1. Fibrinogen + Thrombin – Fibrin Clot
- 2. t-PA + Plasminogen – Plasmin
- 3. Plasmin + Fibrin – Degradation products of Fibrin
  - Fibrinolysis is strictly dependent on t-PA concentration
- Manual Version:
  - fibrin clot in reagent tube; clot lysis endpoint: release of air bubbles from the clot
- Automated Version:
  - fibrin clot in rotor cuvettes; measurement of absorption at 340 nm; clot lysis endpoint: when a threshold absorption level is reached
- Utility:
  - in vitro potency test
  - stability assessment: proteolytic activity and fibrin binding
  - quantification on t-PA
- Advantage: sensitive, accurate, precise and automated

In-vitro thrombolytic Assay

- Lysis of human blood clots in human plasma
  - measurement of decrease of blood clot weight as a function of time
  - determination of blood clot half life
- t-PA is clearly limiting the reaction in plasma, linear response to the t-PA concentration added
- Utility:
  - measurement of thrombolytic activity of t-PA in plasma
- Advantage: in-vitro assay with good relation to in-vivo activity
- Disadvantage: not automated, limited capacity

Fig. 8: In-vitro assay systems for the measurement of the proteolytic, fibrinolytic and thrombolytic potency of t-PA

a fibrin clot is formed; 2. Plasminogen and t-PA bind to fibrin, t-PA cleaves plasminogen into plasmin; 3. Plasmin cleaves fibrin into degradation products, the clot is resolved. This clot lysis assay can be performed in two versions. In the manual version the fibrin clot is formed in a reagent tube and air bubbles are included by shaking the reagent mixture before the clot forms. The endpoint of the lysis reaction is reached when all air bubbles are released from the fibrin clot. In the automated version of the clot lysis assay the clot is formed in a cuvette and the lysis reaction is monitored by measuring the absorbance change at 340 nm.

The clot lysis assay for measuring the fibrinolytic activity of t-PA is an example for a complex biochemical assay which can be carried out reproducibly, accurate, precise, sensitive and automated. Assay variation is very low, with coefficients of variation usually below 5 percent. The potency of t-PA samples is always measured against a reference standard with well defined biological activity.

The described in-vitro clot lysis assay mimics the mechanism of action of t-PA in-vivo. By variation of the sample principle, assay modifications have been developed which are even closer to the in-vivo situation of fibrinolysis. Instead of purified preparations of plasminogen and fibrinogen, human plasma can be used. The measurement of lysis of the plasma clot can be performed in the same way as for purified compounds.

The fibrinolytic assay system which is closest to the in-vivo situation, called "hanging clot" assay, uses whole human blood clots which are lysed in human plasma [12]. T-PA is the limiting substance in the fibrinolytic reaction cascade; without addition of t-PA to the plasma no lysis occurs. In this system the fibrinolytic activity is determined by plotting the kinetic for the loss in blood clot weight versus the concentration of t-PA in the reaction mixture. Due to the fact that the hanging clot assay is very time-consuming and not automated, its application is limited to specific investigations and it is not used as a stability indicating assay. A general requirement for a stability test is a high capacity and throughput of samples, as well as a high accuracy and precision. Both criteria are fulfilled by the enzymatic in-vitro clot lysis assay for t-PA.

For many protein drugs no enzymatic activity can be measured. This is normally the case for monoclonal antibodies which bind to a certain antigen, for soluble receptors or receptor ligands and for virus subunit proteins. In these cases, a choice for an in-vitro potency asasy can be a competitive specific binding assay (CSBA). The principle of the CSBA test is shown in Fig. 9 for one monoclonal antibody product. The competition assay can be designed as a sandwich assay where the soluble antigen is fixed to microtiter plates. The displacement of a conjugated form of the monoclonal antibody is measured for the test sample and for reference material of the antibody within the same assay. From the displacement calibration curve for reference material a relative binding potency for the sample can be determined. This assay format can be performed with a high accuracy, precision and capacity in sample throughput.

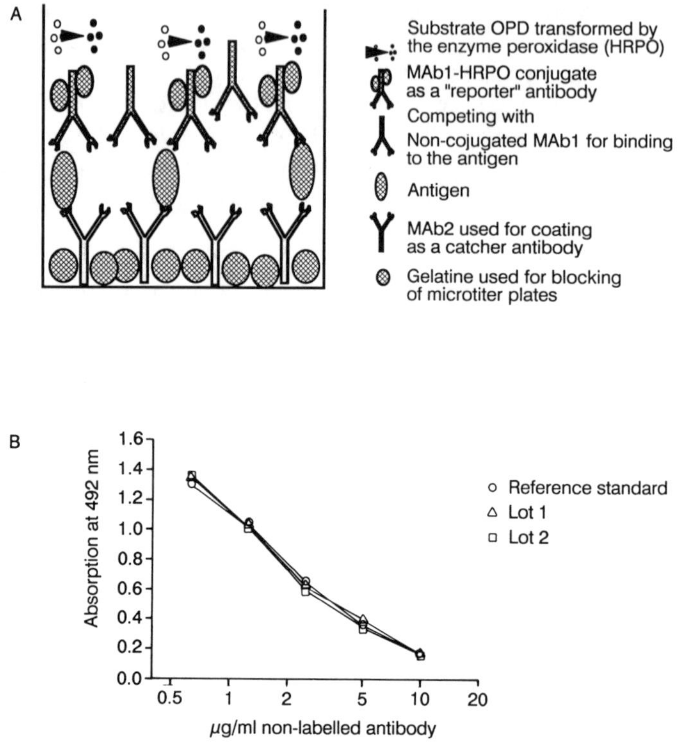

Fig. 9: Competitive Specific Binding Assay for the measurement of binding potency of a monoclonal antibody
(A) Scheme of the assay
(B) Calibration curves of standard and test samples

However, a specific binding assay does not necessarily reflect the "true potency" of a protein drug. Therefore, care has to be taken to correlate the data from in-vitro binding assay with more relevant biological test systems. In addition, the stability indicating properties of such an assay has to be established (Fig. 10).

A different approach for measuring stability has to be chosen for biopharmaceuticals where no enzymatic activity and no binding activity can be measured. For erythropoetin a specific monoclonal antibody was developed which detects only denatured protein. The product is bound to microtiter plates with the aid of polyclonal antibodies and is then probed with this specific monoclonal antibody. This antibody can only bind if denatured product is available. Using denatured product as a reference standard in the assay, the amount of denatured protein in the product can be determined. This type of an assay has been

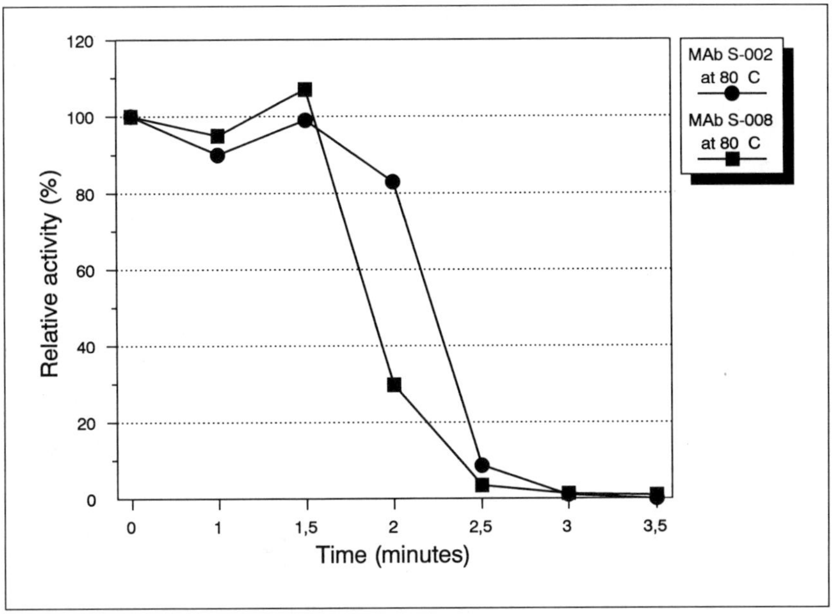

Fig. 10: Stability indicating properties of the Competitive Specific Binding Assay CSBA. Decrease of binding potency of two monoclonal antibodies after incubation at +80 °C

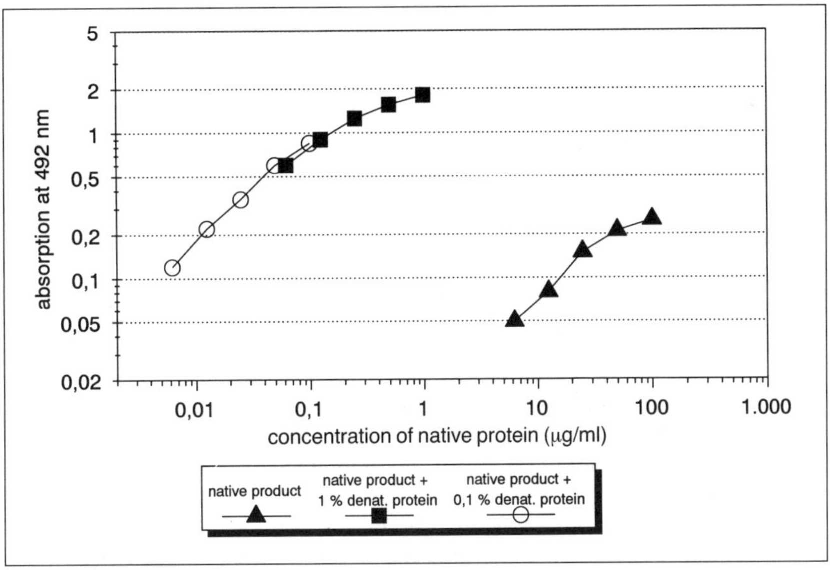

Fig. 11: Detection of denatured product with a specific ELISA-test, based on a specific monoclonal antibody which binds only to denatured product

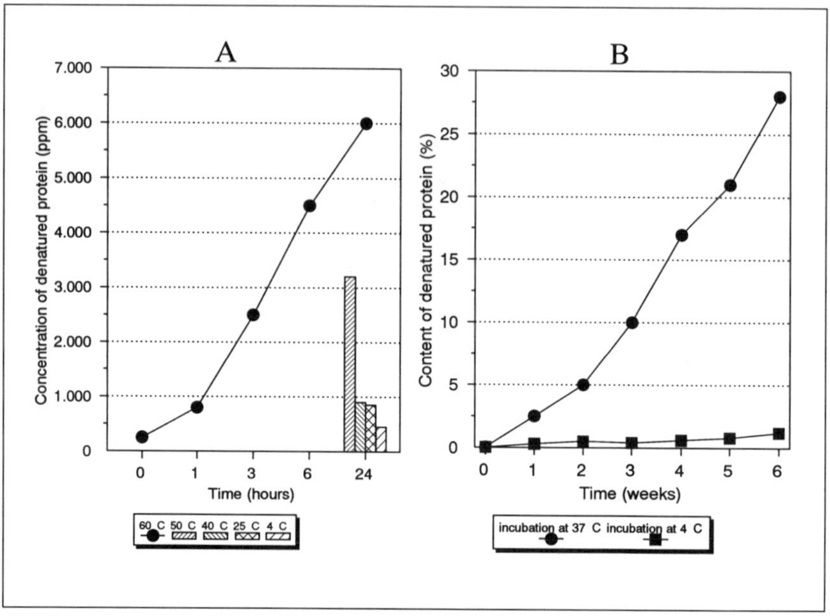

Fig. 12: Stability indicating properties of the ELISA-test for denatured product
(A) Detection of denatured protein after incubation at different temperatures (short term)
(B) Detection of denatured protein after incubation at 37 °C and at 4 °C (long term)

proven to be very sensitive. The product contains normally less than 0.1% denatured product. Spikes of 0.1% and 1% denatured protein in the assay can be detected (Fig. 11). The stability indicating properties of this assay have been established (Fig. 12A and 12B).

6. Choice of the appropriate final formulation and dosage and their importance for protein stability

Proteins are purified by chromatographic and ultrafiltration techniques, thereby never leaving the liquid phase. The experience that has been made with any specific protein during process development is the basis for the development of an appropriate formulation. The final step in purification is often identical with the formulation of the protein.

The pI-value of a protein, determined by isoelectric focussing or chromatofocussing, is an important information for the development of a purification strategy which includes ion-exchange chromatography. At the same time the

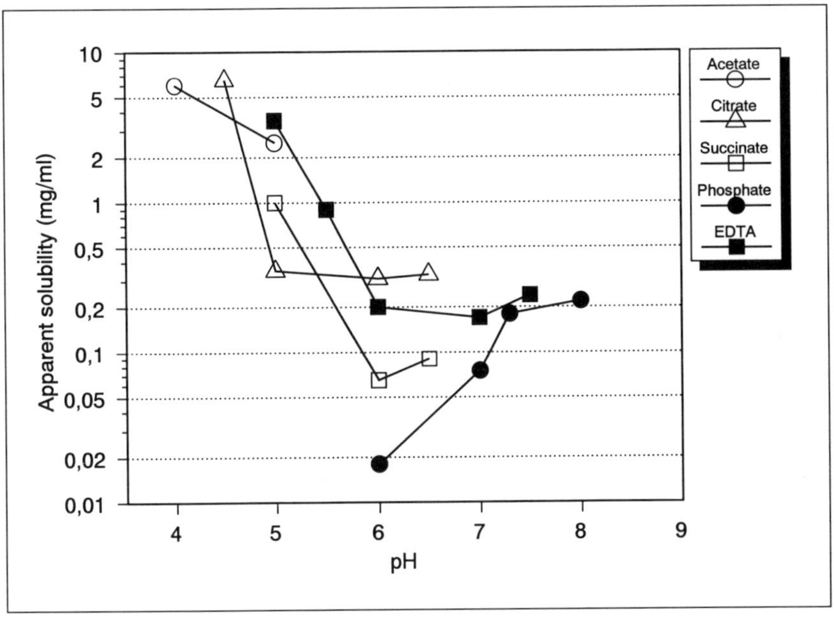

Fig. 13: Solubility of t-PA in different buffer systems (20 mM, at 25 °C) as a function of pH

pI-value or pI-range is of relevance for the choice of the pH of the formulation buffer, because the pH influences solubility and stability of the protein preparation. An example for the solubility of t-PA in different buffer systems as a function of pH is shown in Fig. 13. The buffer system itself and the pH within a defined buffer system have great impact on the solubility of the t-PA product. The salt concentration and the ionic strength do also influence the solubility of t-PA (Fig. 14).

For the development of a stable protein formulation the buffer conditions have to be investigated very carefully using different stability indicating assays. The t-PA stability in solution was examined, depending on the pH of the formulation. The decrease in monomer content is critical at alcaline pH (at pH 8 and pH 9) (Fig. 15). The content of single chain t-PA is constant at pH 4 and 5, and decreases with increasing pH. The rate for transformation of the single-chain form to the two-chain form is highest at pH 9 (Fig. 16). The potency of t-PA measured in an in-vitro clot lysis assay is most stable at pH 5 and pH 6 and decreases gradually at higher and lower pH ranges (Fig. 17).

Avoidance of aggregate formation is one of the most critical issues during development of a stable formulation. Accelerated stability studies can be useful to shorten the response time for aggregation problems with specific formulations. An example is given for three different monoclonal antibodies in Fig. 18,

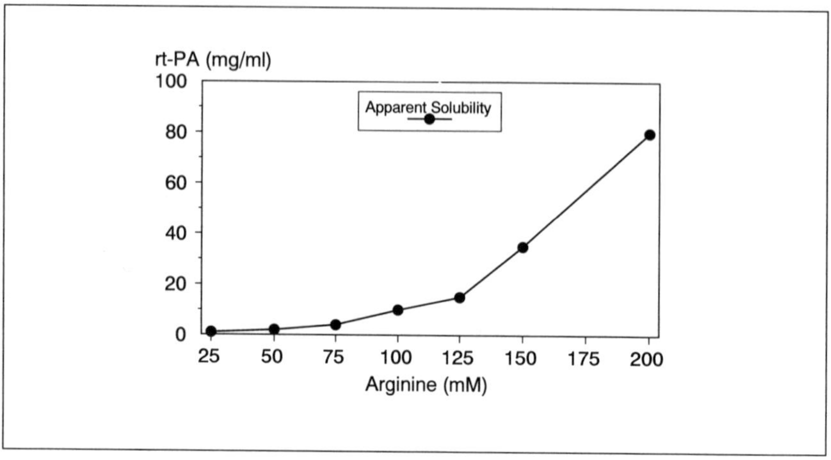

Fig. 14: Solubility of t-PA as a function of salt (arginine phosphate) concentration

where the formation of a small amount of non-soluble aggregates in two different formulations occured. This protein aggregation could be detected by UV-measurement at 320 nm. From the results at higher storage temperature (25 °C and 37 °C) it can be predicted for a given formulation after a short storage period, whether aggregation will cause problems.

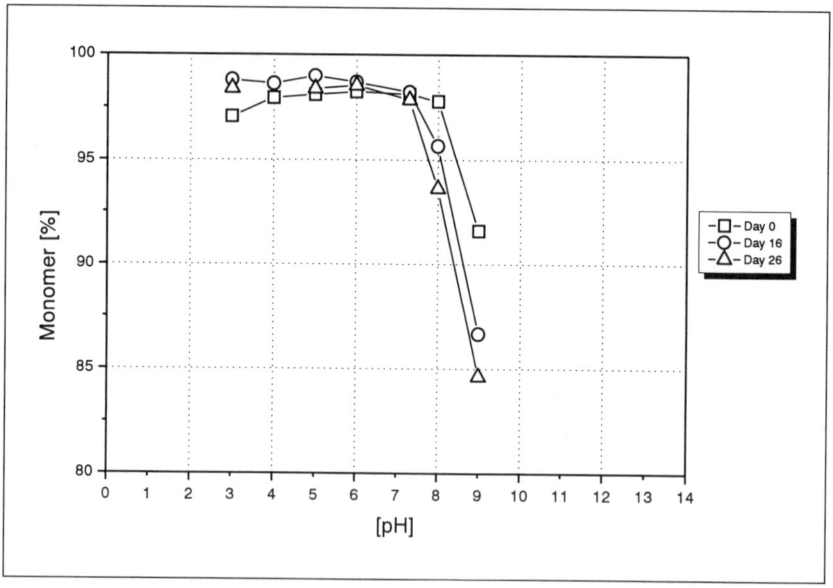

Fig. 15: Stability of t-PA at 25 °C in solution. Monomer content as function of pH

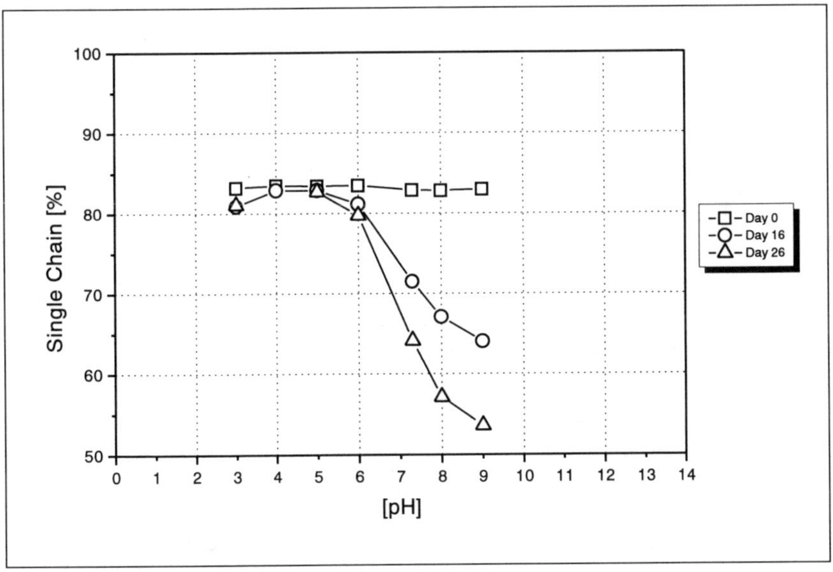

Fig. 16: Stability of t-PA at 25 °C in solution. Content of single-chain t-PA as a function of pH

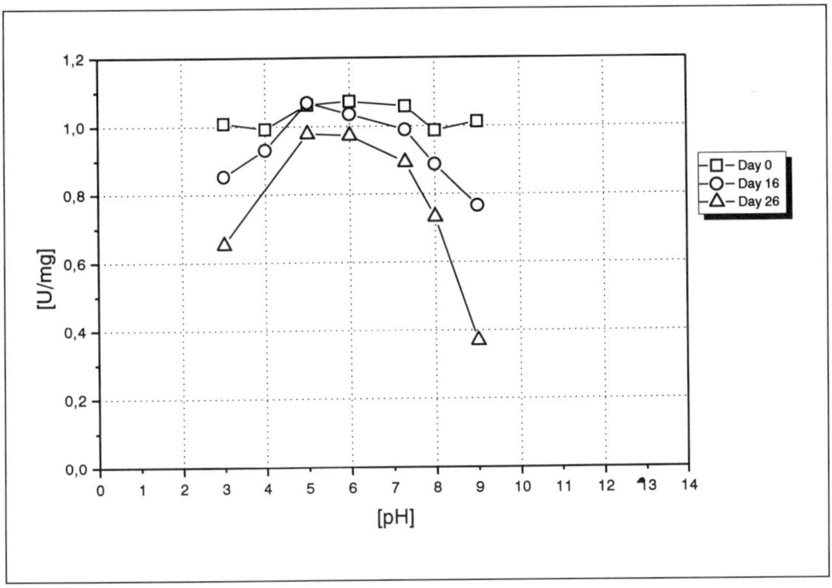

Fig. 17: Stability of t-PA at 25 °C in solution. Specific activity of t-PA as a function of pH

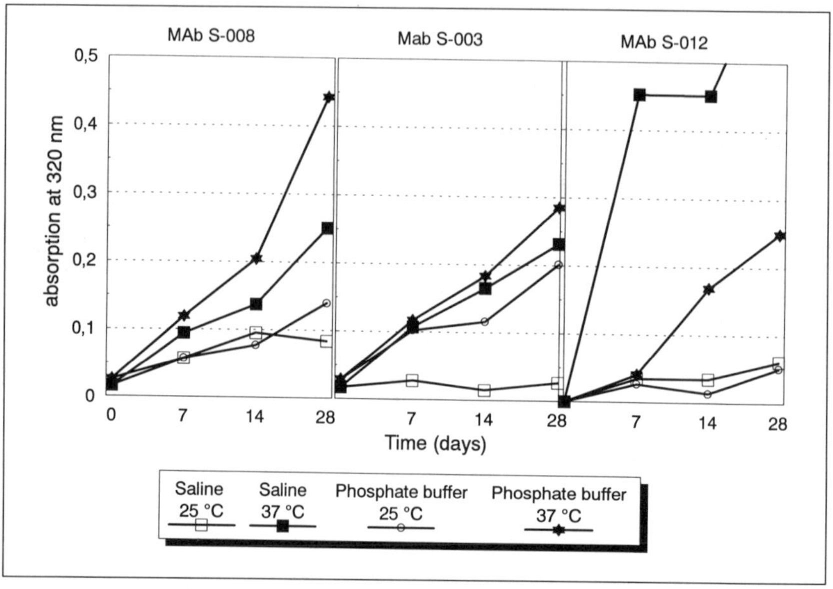

Fig. 18: Detection of aggregate formation for 3 different antibodies in 2 different buffer systems
The antibody solutions were incubated at 25 °C and 37 °C. Aggregate formation was measured by UV-absorption at 320 nm

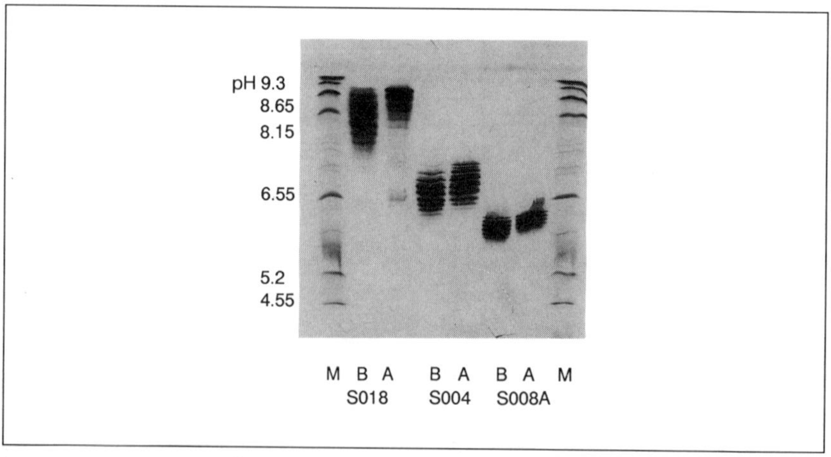

Fig. 19: Deamidation of Monoclonal Antibodies analysed by Isoelectric Focussing
M: PI marker proteins
A: reference standard
B: monoclonal antibody after storage, 6 months at 25 °C

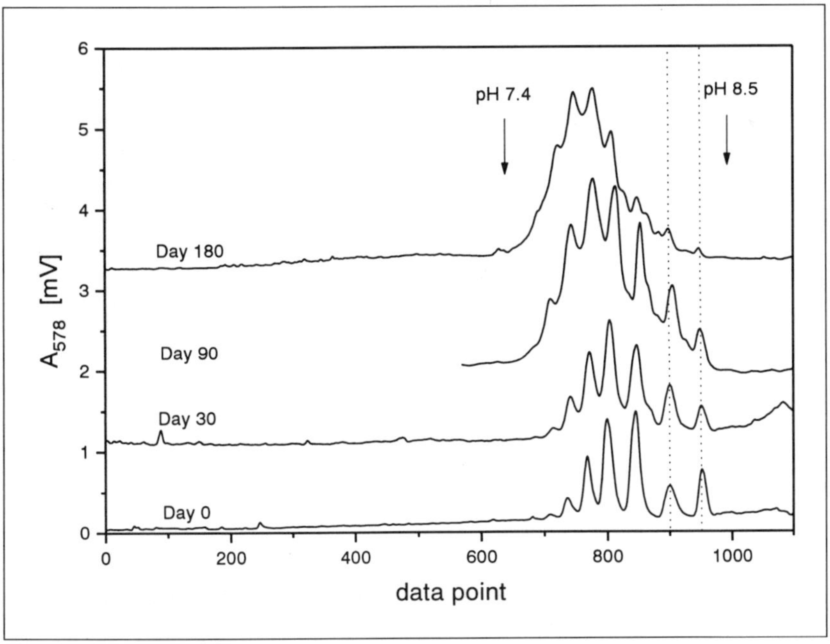

Fig. 20: Deamidation of a monoclonal antibody after incubation in solution, measured by densitometric scanning of IEF gels

For monoclonal antibodies in liquid formulation deamidation is one of the most often observed stability problems. Deamidation can be detected by isoelectric focussing and quantified by densitometric scanning of the gel. The deamidation process is mainly dependent on the storage temperature, on the composition and on the pH of the formulation buffer.

Examples for deamidation are shown for 3 different monoclonal antibodies in Fig. 19. The deamidation process is accelerated at higher pH. Deamidation can be measured by densitometric scanning of IEF gels. An example is given in Fig. 20 for a monoclonal antibody, where deamidation occured during a time period of 6 month.

7. Examples for stable and nonstable biopharmaceuticals

Based on a careful formulation development, an acceptable long-term stability of biopharmaceuticals can be achieved. However, even when formulation development leads to promising results, only real time stability data with the final formulation in a defined final dosage container can give the certainty about product stability.

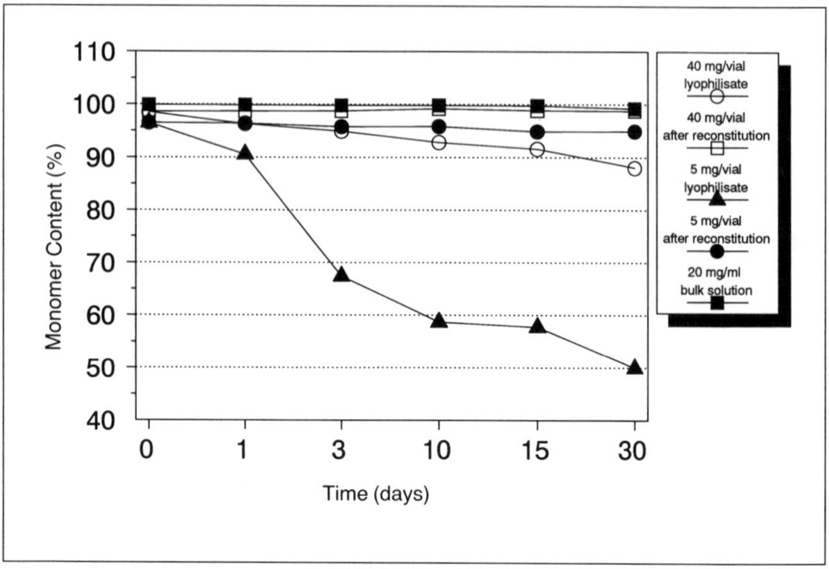

Fig. 21: Stability of a monoclonal antibody at 37 °C in different formulations: Decrease in monomer content as a function of time

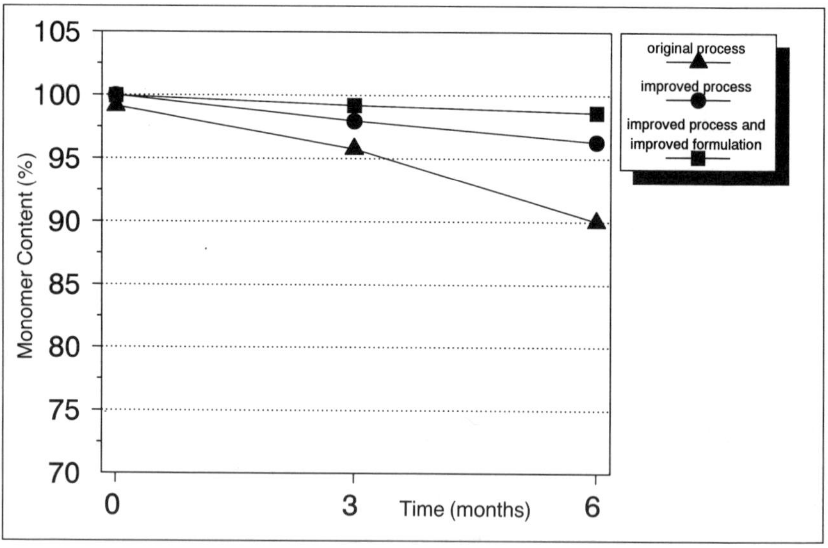

Fig. 22: Stability of a lyophilized preparation of a monoclonal antibody at 25 °C. Influence of lyophilization process conditions and formulation on aggregate formation of the product

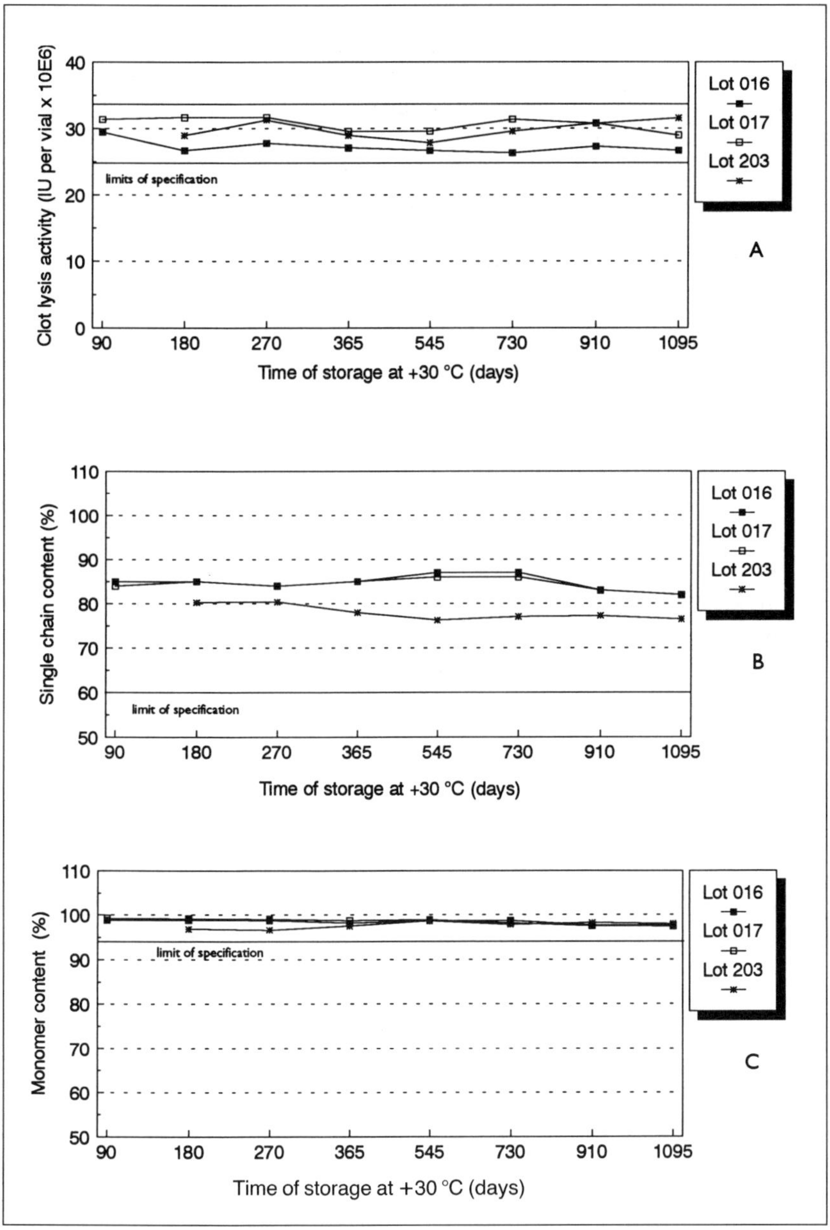

Fig. 23: Stability of 3 different lots of t-PA (Actilyse), stored for 3 years at + 30 °C. Lyophilized final product, 50 mg per vial
(A) specific fibrinolytic activity of t-PA
(B) t-PA single chain content
(C) t-PA Monomer content

For example, a lyophilized formulation was developed for a monoclonal antibody. This formulation led during the experimental stage to an acceptable stability at low temperature (2–8 °C). During clinical trials the dose had to be changed to a lower protein concentration in an otherwise unchanged formulation. With the decrease in protein content in the lyophilisate the tendency to aggregate formation increased dramatically (Fig. 21). Interestingly, with respect to aggregation, the product was much more stable in liquid solution than in the lyophilisate. However, by improvement of the lyophilization process a much better stability of the final product was obtained (Fig. 22). After improvement of the lyophilization process and the formulation, the product was stable for 6 months at 25 °C. In the meantime also a liquid formulation was developed, which lead to a product stability at 2–8 °C for more than 2 years.

For t-PA a stable lyophilized formulation was chosen, due to its lower stability in solution. In this lyophilized formulation the product is stable for more than 3 years at temperatures up to 30 °C (Fig. 23). The specific activity remains constant over a time period of 3 years at a temperature of +30 °C (Fig. 23A). There is also no decrease in single-chain content (Fig. 23B) and in the monomer content of the product (Fig. 23C). During a time period of 3 years at +30 °C no degradation products are detectable by SDS-PAGE (Fig. 24) and the protein banding pattern in the IEF-gel remains constant (Fig. 25). These results demonstrate that very stable protein formulations can be developed, which contain no additional protein as stabilizer.

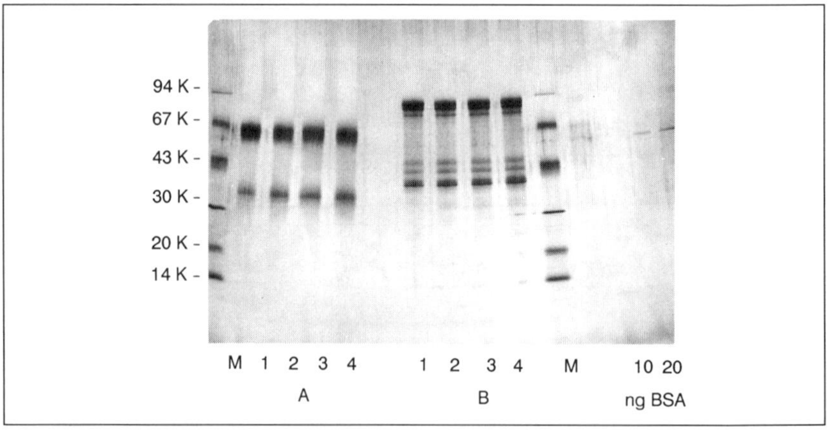

Fig. 24: SDS-PAGE for stability testing of t-PA
M: molecular weight marker proteins
1: reference standard
2: lot 0016 stored for 1 year at +30 °C
3: lot 0016 stored for 2 years at +30 °C
4: lot 0016 stored for 3 years at +30 °C
A) reduced samples
B) reduced and carboxymethylated samples

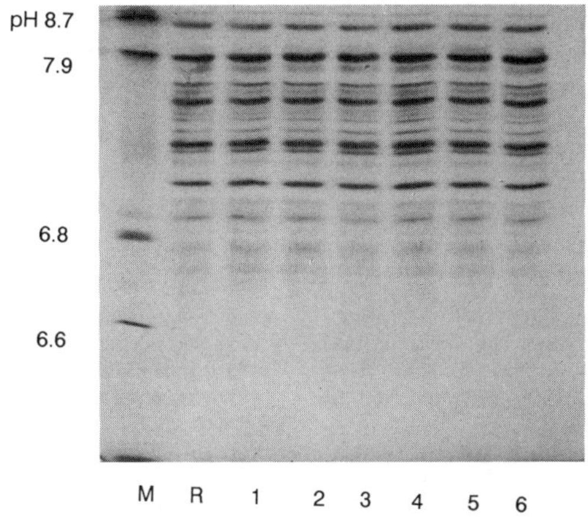

Fig. 25: IEF for stability testing of t-PA
M: PI-marker proteins
R: reference standard
1: lot 0016, 3 years at −70 °C
2: lot 0016, 3 years at +30 °C
3: lot 0017, 3 years at −70 °C
4: lot 0017, 3 years at +30 °C
5: lot 203, 2.5 years at −70 °C
6: lot 203, 2.5 years at +30 °C

Conclusions

Due to the complexity and heterogeneity of protein structures, multiple analytical methods for stability testing must be employed. Proteins are sensitive to chemical modification and to denaturation or aggregation processes. The conformation of the protein is important for the activity and the non-antigenicity of the protein containing pharmaceutical. Therefore, in addition to an extensive analytical product characterization, a stable and reproducible manufacturing process is required. A defined and validated manufacturing process is the basis for lot to lot consistency and a predictable shelf-life of the biopharmaceutical.

Stability indicating test methods must be validated to be suitable to detect potential degradation products or modified forms. Potency testing for biopharmaceuticals depends very much on the product characteristics and the nature of the individual protein drug. In many cases biomimic in-vitro assays can be developed, which are of relevance for the in-vivo efficacy and can be performed in an automated form with high accuracy and precision.

The development and choice of the formulation is of outstanding importance for the stability of biopharmaceuticals. Accelerated stability studies can provide an important early information concerning protein stability in different formulations. However, the shelf-life of a protein containing pharmaceutical can only be defined based on real time stability data in the final formulation and dosage form. Data for this paper demonstrate that formulations for biopharmaceuticals can be developed, which support a shelf-life of more than 3 years at temperatures up to +30 °C.

References

[1] Werner, R.-G., New Biotechnology as a Logical Outgrowth of Molecular Biology, Arzneimittel-Forschung/Drug Research **36**, 9, 1429–1436 (1986)

[2] Werner, R.-G., F. Walz, W. Noé, A. Konrad, Safety and Economic Aspects of Continous Mammalian Cell Culture, Journal of Biotechnology, **22**, 51–68 (1992)

[3] Werner, R. G., C.-H. Pommer, Sucessful Development of Recombinant DNA-derived Pharmaceuticals, Arzneimittel-Forschung/ Drug Research, **40** (II) 11, 1274–1283 (1990)

[4] Werner, R.-.G., Purification of Proteins: Converting a Culture Broth into a Medicine Elsevier Science Publishers, 3–22 (1991)

[5] Grimm, W., (ed), Stability of Drug Products, APV Paperback, Band 16, Wissenschaftliche Verlagsgesellschaft mbH Stuttgart (1987)

[6] Commission of EC, A Guideline for the Stability Testing of New Active Drug Substances and Products (Draft) 10th December 10 pages (1991)

[7] Vehar, G. A., M. W. Spellman, B. A. Keyt, C. K. Ferguson, R. G. Keck, R. C. Chlonpek, R. Harris, W. F. Bennet, S. E. Builder, W. S. Hancock, Characterization Studies of Human Tissue-Type Plasminogen Activator Produced by Recombinant DNA Technology, Cold Spring Harbor Symp. Quant. Biol. **51**, 551–562 (1986)

[8] Sobel, B. E., D. Collen, E. B. Grossbard (Eds.), Tissue Plasminogen Activator in Thrombolytic Therapy, New York: Marcel Dekker (1987)

[9] Seifried, E., K. Lechner, H. Heimpel (Eds), Neuere Entwicklungen in der fibrinolytischen Therapie, Klin. Wochenschr. 66 (Suppl. XII), 1–149 (1988)

[10] Werner, R.-G., H. Langlouis-Gau, F. Walz, H. Allgaier, H. Hoffmann, Validation of Biotechnological Production Processes, Arzneimittel-Forschung/Drug Research, **38** (I), 6. 855–862 (1988)

[11] Werner, R.-G., H. Hoffmann, Biotechnische Produktion einer neuen Generation von Arzneimitteln: Therapie mit körpereigenen Wirkstoffen, Praxis der Naturwissenschaften Chemie Heft 5/38, 15. Juli (1989)

[12] Werner, R.-G., S. Bassarab, H. Hoffmann, M. Schlüter, Quality Aspects of Fibrinolytic Agents Based on Biochemical Characterization, Arzneimittel-Forschung/Drug Research **41** (II), 11, 1196–1200 (1991)